Be prepared...
To learn...
To succeed...

Get **REA**dy. It all starts here. REA's preparation for the MEAP is **fully aligned** with the Grade Level Content Expectations of the Michigan Department of Education.

Visit us online at
www.rea.com

READY, SET, GO!

Michigan
MEAP
Grade 8
English Language Arts

Staff of Research & Education Association

Research & Education Association
Visit our website at
www.rea.com

The standards and benchmarks presented in this book were created and implemented by the Michigan Department of Education (DOE). For further information, visit the Michigan DOE website at *www.michigan.gov.mde*.

Research & Education Association
61 Ethel Road West
Piscataway, New Jersey 08854
E-mail: info@rea.com

Ready, Set, Go!
MEAP English Language Arts, Grade 8

Printed in the United States of America

Library of Congress Control Number 2006924966

International Standard Book Number 0-7386-0100-4

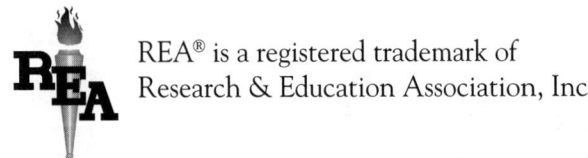

REA® is a registered trademark of
Research & Education Association, Inc.

TABLE OF CONTENTS

About Research & Education Association

Founded in 1959, Research & Education Association is dedicated to publishing the finest and most effective educational materials—including software, study guides, and test preps—for students in middle school, high school, college, graduate school, and beyond. Today, REA's wide-ranging catalog is a leading resource for teachers, students, and professionals.

We invite you to visit us at *www.rea.com* to find out how "REA is making the world smarter."

Acknowledgments

We would like to thank Larry B. Kling, Vice President, Editorial, for his editorial direction; Pam Weston, Vice President, Publishing, for setting the quality standards for production integrity and managing the publication to completion; Christine Reilley, Senior Editor, for project management; Diane Goldschmidt, Associate Editor, for post-production quality assurance; Christine Saul, Senior Graphic Artist, for cover design; Jeremy Rech, Graphic Artist, for interior page design; and Jeff LoBalbo, Senior Graphic Artist, for post-production file mapping.

We also gratefully acknowledge the writers, educators, and editors of REA, Northeast Editing, and Publication Services for content development, editorial guidance, and final review.

SUCCEEDING ON THE MEAP—
Grade 8 English Language Arts

ABOUT THIS BOOK

Our book provides excellent preparation for the Michigan Educational Assessment Program (MEAP)—Grade 8 ELA. Inside you will find exercises that are designed to provide students with the instruction, practice, and strategies needed to do well on this achievement test.

This book is divided into several parts: a **pretest** modeled after the MEAP. The pretest includes these sections:

Part 1: Reading
　　Part 1A: Paired Reading Selections (40 minutes)
　　Part 1B: Response to the Paired Reading Selections (25 minutes)

Part 2: Reading: Independent Reading Selections (30 minutes)

Part 3: Reading: Passage Linked to Previous Grade (30–50 minutes)*

Part 4: Writing
　　Part 4A: Writing from Knowledge and Experience (50 minutes)
　　Part 4B: Student Writing Sample (30 minutes)

Following the pretest is a lesson section, which teaches students about the different types of MEAP questions and skills on the reading and writing portions of the test. Finally, this book includes a full-length posttest.

Begin by assigning students the pretest. Answers and answer explanations follow the pretest. Then work through each of the lessons one by one. When students have completed the book, they should complete the posttest. Answers and answer explanations follow the posttest.

HOW TO USE THIS BOOK

FOR STUDENTS: To make getting through the book as easy as possible, we've included icons shown on the next page that highlight sections like lessons, questions, and answers. You'll find that our practice tests are very much like the actual MEAP you'll encounter on test day. The best way to prepare for a test is to practice, so we've included drills with answers throughout the book, and our two practice tests include detailed answers.

*The time for Part 3 varies depending on the form number used.

FOR PARENTS: Michigan has created grade-appropriate learning standards that are listed in the table in this introduction. Students need to meet these objectives as measured by the MEAP. Our book will help your child review for the MEAP and prepare for the exam. It includes review sections, drills, and two practice tests complete with explanations to help your child focus on the areas he/she needs to work on to help master the test.

FOR TEACHERS: No doubt, you are already familiar with the MEAP and its format. Begin by assigning students the pretest. An answer key and detailed explanations follow the pretest. Then work through each of the lessons in succession. When students have completed the subject review, they should move on to the posttest. Answers and answer explanations follow the posttest.

ICONS EXPLAINED

Icons make navigating through the book easier by highlighting sections like lessons, questions, and answers as explained below:

 Question

 Answer

 Tip

 Lesson

 Activity

 Writing Task

WHY STUDENTS ARE REQUIRED TO TAKE THE MEAP

The Michigan Educational Assessment Program (MEAP) was initiated by the State Board of Education in 1969. Since 2003, the MEAP ELA assessments have reflected the ELA Content Standards and Benchmarks as described in the 1995 Michigan Curriculum Framework. The 1995 standards, however, were written to grade-level clusters and not to any particular grade level. The Fall 2005 MEAP English Language Arts (ELA) assessment has undergone changes to better reflect the requirements of the No Child Left Behind Act of 2001 (NCLB), as well as the 2004 Michigan ELA Grade Level Content Expectations (GLCEs). The reading portion of the test now contains:

- One narrative passage

- One informational passage

- Two additional narrative and/or informational passages

- One passage linked to previous grade

- Approximately eight multiple-choice questions per passage

- Five multiple-choice cross-text items

- One constructed-response item

The writing portion of the test now contains the following sections:

- Writing from Knowledge and Experience—Students are given two pages maximum (front and back) for this response, which requires them to write an essay based on a writing prompt. This response must be written in one testing session and will be scored as first-draft writing using a six-point rubric.

- Peer Response to a Student Writing Sample—Students will be allowed a one-page maximum response to a direct question, such as a critical response, to a piece of student writing. This response is scored as first-draft writing using a four-point rubric.

- Multiple-Choice Items Based on Student Writing Sample—Students will be asked revision and editing questions that refer back to the same student writing sample that is used for the Peer Response.

The following standards and benchmarks are tested on the Grade 8 MEAP ELA.

MEAP READING STANDARDS/BENCHMARKS, GRADE 8*

Word Study

		Page Numbers
R.WS.08.01	Use word structure, sentence structure, and prediction to aid in decoding and understanding the meanings of words encountered in context.	67
R.WS.08.02	Use structural, syntactic, and semantic analysis to recognize unfamiliar words in context (idioms, analogies, metaphors, and similes to infer the history of the English language, common word origins, or syllabication).	67
R.WS.08.04	Know the meaning of frequently encountered words in written and oral contexts (research to support specific words).	67
R.WS.08.07	Use strategies (e.g., prior knowledge, text features, structures) and authentic content-related resources to determine the meaning of words and phrases in context (e.g., historical terms, content area vocabulary, literary terms).	67

Narrative Text

		Page Numbers
R.NT.08.01	Investigate through classic and contemporary literature recognized for quality and literary merit various examples of distortion and stereotypes such as those associated with gender, race, culture, age, class, religion, and other individual differences.	90
R.NT.08.02	Analyze elements and style of narrative genres (e.g., historical fiction, science fiction, realistic fiction).	90
R.NT.08.03	Analyze the role of rising and falling actions, minor characters in relation to conflict, and credibility of the narrator.	90
R.NT.08.04	Analyze how authors use symbolism, imagery, and consistency to develop credible narrators, rising and falling actions, and minor characters.	90

* The standards and benchmarks presented in this table were created and implemented by the Michigan Department of Education (DOE). For further information, visit the Michigan DOE at *http://www.michigan.gov.mde*.

Informational Text

		Page Numbers
R.IT.08.01	Analyze elements and style of informational genre (e.g., comparative essays, newspaper writing, technical writing, persuasive essay).	110
R.IT.08.02	Analyze organizational patterns (e.g, theory, evidence, sequence).	110
R.IT.08.03	Explain how authors use text features to enhance the understanding of central, key, and supporting ideas (e.g., illustrations, author's pages, prefaces, marginal notes).	110

Critical Standards

		Page Numbers
R.CS.08.01	Evaluate the appropriateness of shared, individual, and expert standards based on purpose, context, and audience in order to assess their own work and work of others.	110

Comprehension

		Page Numbers
R.CM.08.01	Connect personal knowledge, experience, and understanding of the world to themes and perspectives in the text.	129
R.CM.08.02	Read, retell, and summarize grade-level appropriate narrative and informational texts.	129
R.CM.08.03	State global themes, universal truths, and principles within and across texts to create a deeper understanding.	129
R.CM.08.04	Apply significant knowledge from what has been read in grade-level appropriate science and social studies texts.	129

Writing Genres

		Page Numbers
W.GN.08.01	Write a cohesive, narrative piece that includes appropriate conventions for the genre (e.g., historical fiction, science fiction, realistic fiction) and employ literary and plot devices (e.g., narrator credibility, rising and falling action, and/or conflict, transitional language, and imagery).	151
W.GN.08.02	Write a historical expository piece (e.g., journal, biography, simulated memoir) that includes appropriate organization, illustrations, marginal notes, and/or annotations.	151

Writing Process

Personal Style

Grammar and Usage

Spelling

Handwriting

TIPS FOR THE STUDENT

Students can do plenty of things before and during the actual test to improve their test-taking performance. The good thing is that most of the tips described in the following pages are easy!

Preparing for the Test

Test Anxiety

Do you get nervous when your teacher talks about taking a test? A certain amount of anxiety is normal and it actually may help you prepare better for the test by getting you motivated. But too much anxiety is a bad thing and may keep you from properly preparing for the test. Here are some things to consider that may help relieve test anxiety:

- Share how you are feeling with your parents and your teachers. They may have ways of helping you deal with how you are feeling.

- Keep on top of your game. Are you behind in your homework and class assignments? A lot of your classwork-related anxiety and stress will simply go away if you keep up with your homework assignments and classwork. And then you can focus on the test with a clearer mind.

- Relax. Take a deep breath or two. You should do this especially if you get anxious while taking the test.

Study Tips & Taking the Test

- **Learn the Test's Format.** Don't be surprised. By taking a practice test ahead of time you'll know what the test looks like, how much time you will have, how many questions there are, and what kinds of questions are going to appear on it. Knowing ahead of time is much better than being surprised.

- **Read the Entire Question.** Pay attention to what kind of answer a question or word problem is looking for. Reread the question if it does not make sense to you, and try to note the parts of the question needed for figuring out the right answer.

- **Read All the Answers.** On a multiple-choice test, the right answer could also be the last answer. You won't know unless you read all the possible answers to a question.

- **It's Not a Guessing Game.** If you don't know the answer to a question, don't make an uneducated guess. And don't randomly pick just any answer either. As you read over each possible answer to a question, note any answers which are obviously wrong. Each obviously wrong answer you identify and eliminate greatly improves your chances at selecting the right answer.

- **Don't Get Stuck on Questions.** Don't spend too much time on any one question. Doing this takes away time from the other questions. Work on the easier questions first. Skip the really hard questions and come back to them if there is still enough time.

- **Accuracy Counts.** Make sure you record your answer in the correct space on your answer sheet. Fixing mistakes only takes time away from you.

- **Finished Early?** Use this time wisely and double-check your answers.

Sound Advice for Test Day

The Night Before. Getting a good night's rest keeps your mind sharp and focused for the test.

The Morning of the Test. Have a good breakfast. Dress in comfortable clothes. Keep in mind that you don't want to be too hot or too cold while taking the test. Get to school on time. Give yourself time to gather your thoughts and calm down before the test begins.

Three Steps for Taking the Test

1) **READ.** Read the entire question and then read all the possible answers.

2) **ANSWER.** Answer the easier questions first and then go back to the more difficult questions.

3) **DOUBLE-CHECK.** Go back and check your work if time permits.

TIPS FOR PARENTS

- Encourage your child to take responsibility for homework and class assignments. Help your child create a study schedule. Mark the test's date on a family calendar as a reminder for both of you.

- Talk to your child's teachers. Ask them for progress reports on an ongoing basis.

- Commend your child's study and test successes. Praise your child for successfully following a study schedule, for doing homework, and for any work done well.

- Test Anxiety. Your child may experience nervousness or anxiety about the test. You may even be anxious, too. Here are some helpful tips on dealing with a child's test anxiety:

  - Talk about the test openly and positively with your child. An ongoing dialogue not only can relieve your child's anxieties but also serves as a progress report of how your child feels about the test.

  - Form realistic expectations of your child's testing abilities.

  - Be a "Test Cheerleader." Your encouragement to do his or her best on the test can alleviate your child's test anxiety.

PRETEST

GENERAL DIRECTIONS:

This pretest is divided into three reading and two writing sections. For the reading sections, read each of the passages and then, on your Answer Sheet, answer the questions that follow.

PART 1—READING

PART 1A: PAIRED READING SELECTIONS

DIRECTIONS:

In Part 1A, you will read two selections and answer the multiple-choice questions that follow each selection. You will then answer some questions that will ask you to think about both of the selections. You may look back at these two selections as often as needed during Part 1.

You may underline, highlight, or write notes in this booklet to help you, but you must mark all of your answers in Part 1A of your **Answer Sheets** on page 45.

You may not use any resource materials (dictionaries, grammar books, spelling books, etc.) for any part of this test.

When you have finished Part 1A, STOP.

WAIT. DO NOT GO ON UNTIL TOLD TO DO SO.

DIRECTIONS: Read Selection 1, an excerpt from *Little Women*. Then answer the questions that follow. **You will have 40 minutes to complete this section.**

An Excerpt from *Little Women*
by Louisa May Alcott

We shouldn't enjoy ourselves half so much as we do now. But it does seem so nice to have little suppers and bouquets, and go to parties, and drive home, and read and rest, and not work. It's like other people, you know, and I always envy girls who do such things, I'm so fond of luxury, said Meg, trying to decide which of two shabby gowns was the least shabby.

Well, we can't have it, so don't let us grumble but shoulder our bundles and trudge along as cheerfully as Marmee does. I'm sure Aunt March is a regular Old Man of the Sea to me, but I suppose when I've learned to carry her without complaining, she will tumble off, or get so light that I shan't mind her.

This idea tickled Jo's fancy and put her in good spirits, but Meg didn't brighten, for her burden, consisting of four spoiled children, seemed heavier than ever. She had not heart enough even to make herself pretty as usual by putting on a blue neck ribbon and dressing her hair in the most becoming way.

Where's the use of looking nice, when no one sees me but those cross midgets, and no one cares whether I'm pretty or not? she muttered, shutting her drawer with a jerk. I shall have to toil and moil all my days, with only little bits of fun now and then, and get old and ugly and sour, because I'm poor and can't enjoy my life as other girls do. It's a shame!

So Meg went down, wearing an injured look, and wasn't at all agreeable at breakfast time. Everyone seemed rather out of sorts and inclined to croak.

Beth had a headache and lay on the sofa, trying to comfort herself with the cat and three kittens. Amy was fretting because her lessons were not learned, and she couldn't find her rubbers. Jo would whistle and make a great racket getting ready.

Mrs. March was very busy trying to finish a letter, which must go at once, and Hannah had the grumps, for being up late didn't suit her.

There never was such a cross family! cried Jo, losing her temper when she had upset an inkstand, broken both boot lacings, and sat down upon her hat.

You're the crossest person in it! returned Amy, washing out the sum that was all wrong with the tears that had fallen on her slate.

Beth, if you don't keep these horrid cats down cellar I'll have them drowned, exclaimed Meg angrily as she tried to get rid of the kitten which had scrambled up her back and stuck like a burr just out of reach.

GO ON TO THE NEXT PAGE

Jo laughed, Meg scolded, Beth implored, and Amy wailed because she couldn't remember how much nine times twelve was.

Girls, girls, do be quiet one minute! I must get this off by the early mail, and you drive me distracted with your worry, cried Mrs. March, crossing out the third spoiled sentence in her letter.

There was a momentary lull, broken by Hannah, who stalked in, laid two hot turnovers on the table, and stalked out again. These turnovers were an institution, and the girls called them 'muffs', for they had no others and found the hot pies very comforting to their hands on cold mornings.

Hannah never forgot to make them, no matter how busy or grumpy she might be, for the walk was long and bleak. The poor things got no other lunch and were seldom home before two.

Cuddle your cats and get over your headache, Bethy. Goodbye, Marmee. We are a set of rascals this morning, but we'll come home regular angels. Now then, Meg! And Jo tramped away, feeling that the pilgrims were not setting out as they ought to do.

They always looked back before turning the corner, for their mother was always at the window to nod and smile, and wave her hand to them. Somehow it seemed as if they couldn't have got through the day without that, for whatever their mood might be, the last glimpse of that motherly face was sure to affect them like sunshine.

Questions 1–8

1 What did Meg dislike MOST about her life?

    A   getting headaches so often

    B   watching four children

    C   having only two gowns to choose from

    D   not spending more time having fun

2 With which statement would the author PROBABLY agree?

    A   Girls who have no work to do are very lazy.

    B   Girls who have no work to do are very lucky.

    C   Girls who have a lot of work complain too much.

    D   Girls who have a lot of work should not try to be pretty.

GO ON TO THE NEXT PAGE

3 What is the MAIN idea of this excerpt?

A Having a lot of work makes us appreciate people we love.

B Having people we love helps us get through hard times.

C There is no point in dressing up to baby-sit children.

D All children should respect their parents.

4 The Marches' lives were portrayed by the author as being

A fun and lighthearted.

B boring and dull.

C busy and hectic.

D dreary and sad.

5 The MAIN reason Meg does not put on a blue neck ribbon is because she

A is in a bad mood and is being difficult.

B feels she has no one to look pretty for.

C cannot find the ribbon in the confusion.

D is too busy laughing with Jo.

6 Meg suggests that girls should

A always look nice.

B be quiet.

C work hard.

D not work.

GO ON TO THE NEXT PAGE

7 The author seems to view the girls' mother as

    A   an example of how they should act.

    B   a tyrant who orders them around.

    C   a symbol of how hard they work.

    D   a protector of the girls.

8 Which statement BEST summarizes paragraphs 13 and 14?

    A   Making food can cause people to stop complaining.

    B   When all is not well, small things can cheer people.

    C   Even when we're grumpy, we should be nice to people.

    D   Packing a lunch is important when you'll be gone all day.

GO ON TO THE NEXT PAGE

DIRECTIONS: Read Selection 2, the poem "Love Is Not All." Then answer the questions that follow.

Love Is Not All
by Edna St. Vincent Millay

Love is not all: it is not meat nor drink
Nor slumber nor a roof against the rain;
Nor yet a floating spar to men that sink
And rise and sink and rise and sink again;
Love can not fill the thickened lung with breath,
Nor clean the blood, nor set the fractured bone;
Yet many a man is making friends with death
Even as I speak, for lack of love alone.
It well may be that in a difficult hour,
Pinned down by pain and moaning for release,
Or nagged by want past resolution's power,
I might be driven to sell your love for peace,
Or trade the memory of this night for food.
It well may be. I do not think I would.

Questions 9–16

9 Which of the following BEST describes the person speaking in this poem?

    A clever and loving

    B happy and selfish

    C sad and lonely

    D angry and sullen

10 In the poem "Love Is Not All," Edna St. Vincent Millay contrasts

    A food and hate.

    B love and necessity.

    C friendship and romance.

    D life and death.

GO ON TO THE NEXT PAGE

11 What is the mood of the poem?

    A disappointed

    B serious

    C playful

    D resentful

12 When the poet says "a floating spar to men that sink" (line 3), she is referring to something that

    A brings men together.

    B men fight over.

    C pushes men down into water.

    D helps men stay above water.

13 This poem is MOSTLY about the

    A hardships that prevent us from love.

    B effects of love on survival.

    C speaker's specific love interest.

    D importance of love to human beings.

14 With which of the following statements would the author MOST LIKELY agree?

    A Love can heal physical pain.

    B Some people wouldn't trade love for anything.

    C Love can prevent people from getting hungry.

    D People in love do not feel pain.

GO ON TO THE NEXT PAGE

15 Which line BEST supports the idea that love is NOT the most important thing to the author?

    A   Love can not fill the thickened lung with breath.

    B   Many a man is making friends with death.

    C   I might be driven to sell your love for peace.

    D   I do not think I would.

16 The author would NOT trade her memory of this night for food because the memory

    A   has nothing to do with food.

    B   is not worth much food.

    C   is dearer to her than food.

    D   makes her feel full.

GO ON TO THE NEXT PAGE

CROSS-TEXT QUESTIONS

DIRECTIONS: Questions 17 through 21 ask about *both* of the selections you read. Choose the *best* answer for each question. You may look back at the two selections as often as needed.

17 With which of the following statements would BOTH authors PROBABLY agree?

A We forget about love when we go through hard times.

B A mother's love is the most important thing in the world.

C Love is more important than material possessions.

D Food cannot provide the feeling that love can.

18 Which of the following do BOTH passages include?

A A description of someone who is hurting

B A stated wish for more luxuries

C A cook who is in a bad mood

D A family who is fighting

19 One DIFFERENCE between the women in the passages is that

A the March sisters are poor, but the speaker in the poem is rich.

B the March sisters are healthy, but the speaker in the poem is ill.

C the March sisters live in luxury, but the speaker in the poem lives simply.

D the March sisters are young girls, but the speaker in the poem is mature.

GO ON TO THE NEXT PAGE

20 The perspective of these passages is DIFFERENT in that

   A *Little Women* is written in third person, and "Love Is Not All" is in first person.

   B *Little Women* is written in first person, and "Love Is Not All" is in third person.

   C *Little Women* is written in second person, and "Love Is Not All" is in third person.

   D *Little Women* is written in third person, and "Love Is Not All" is in second person.

21 Which idea is explored in BOTH passages?

   A Kittens scurrying up a person's back

   B People wanting more than they have

   C Men sinking into the ocean

   D A person writing a letter

STOP

PART 1B: RESPONSE TO THE PAIRED READING SELECTIONS

DIRECTIONS:

Write a response to the scenario question that is stated below. Your own ideas and experiences may be used in your response, but you MUST use examples from BOTH reading selections to earn full credit. You may look back at BOTH reading selections at any time. **You will have approximately 25 minutes to complete this part of the test.**

You may write down ideas, organize your thoughts, or write a rough draft in the "Notes/ Planning" section.

Please write your response on pages 46 and 47 of your **Answer Sheets.**

---

22 **Your teacher has asked the students in your class to write an essay on the definition of love. You are ready to begin writing your essay.**

What is love? Why is it important to people? Explain your answer using details from BOTH the excerpt from *Little Women* and the poem "Love Is Not All" to support your answer. Be sure to show how the two reading selections are connected or alike.

Use the Checklist for the Response to the Reading Selections to help you with your response.

WAIT. DO NOT GO ON UNTIL TOLD TO DO SO.

PART 1B: CHECKLIST FOR REVISING AND PROOFREADING

DIRECTIONS:

Use the following checklists as you revise and proofread the writing you have done for Part 1. When you are finished revising, you must write your final copy. Then, proofread your final copy to make sure that all of your revisions have been made.

CHECKLIST FOR REVISION:

_____ Do I have a clear central idea that connects to the theme?

_____ Do I stay focused on the theme?

_____ Do I support my central idea with important details/examples?

_____ Do I need to take out details/examples that DO NOT support my central idea?

_____ Is my writing organized and complete?

_____ Do I use a variety of words, phrases, and/or sentences?

CHECKLIST FOR EDITING:

_____ Have I checked and corrected my spelling to help readers understand my writing?

_____ Have I checked and corrected my punctuation and capitalization to help readers understand my writing?

CHECKLIST FOR PROOFREADING:

_____ Is everything in my final copy the way that I want it?

GO ON TO THE NEXT PAGE

NOTES/PLANNING

GO ON TO THE NEXT PAGE

NOTES/PLANNING

PART 2—READING

INDEPENDENT READING SELECTIONS

DIRECTIONS:

In Part 2, you will read a selection and answer the multiple-choice questions that follow. You may look back at the selection as often as needed in Part 2.

You may underline, highlight, or write notes in this booklet to help you, but you must mark all of your answers in Part 2 of your **Answer Sheets** on page 48.

You may not use any resource materials (dictionaries, grammar books, spelling books, etc.) for any part of this test.

When you have finished Part 2, STOP.

WAIT. DO NOT GO ON UNTIL TOLD TO DO SO.

DIRECTIONS: Read Selection 1, an excerpt from *White Fang*. Then answer the questions that follow. Then you will read Selection 2 and answer the questions that follow. **You will have about 30 minutes to complete this section.**

Excerpt from *White Fang*
by Jack London

They camped early that night. Three dogs could not drag the sled so fast nor for so long hours as could six, and they were showing unmistakable signs of playing out. And the men went early to bed, Bill first seeing to it that the dogs were tied out of gnawing-reach of one another.

But the wolves were growing bolder, and the men were aroused more than once from their sleep. So near did the wolves approach, that the dogs became frantic with terror, and it was necessary to replenish the fire from time to time in order to keep the adventurous marauders at safer distance.

"I've hearn sailors talk of sharks followin' a ship," Bill remarked, as he crawled back into the blankets after one such replenishing of the fire. "Well, them wolves is land sharks. They know their business better'n we do, an' they ain't a-holdin' our trail this way for their health. They're goin' to get us. They're sure goin' to get us, Henry."

"They've half got you a'ready, a-talkin' like that," Henry retorted sharply. "A man's half licked when he says he is. An' you're half eaten from the way you're goin' on about it."

"They've got away with better men than you an' me," Bill answered.

"Oh, shet up your croakin'. You make me all-fired tired."

Henry rolled over angrily on his side, but was surprised that Bill made no similar display of temper. This was not Bill's way, for he was easily angered by sharp words. Henry thought long over it before he went to sleep, and as his eyelids fluttered down and he dozed off, the thought in his mind was: "There's no mistakin' it, Bill's almighty blue. I'll have to cheer him up to-morrow."

. . .

The day began auspiciously. They had lost no dogs during the night, and they swung out upon the trail and into the silence, the darkness, and the cold with spirits that were fairly light. Bill seemed to have forgotten his forebodings of the previous night, and even waxed facetious with the dogs when, at midday, they overturned the sled on a bad piece of trail.

It was an awkward mix-up. The sled was upside down and jammed between a tree-trunk and a huge rock, and they were forced to unharness the dogs in order to straighten out the tangle. The two men were bent over the sled and trying to right it, when Henry observed One Ear sidling away.

GO ON TO THE NEXT PAGE

"Here, you, One Ear!" he cried, straightening up and turning around on the dog.

But One Ear broke into a run across the snow, his traces trailing behind him. And there, out in the snow of their back track, was the she-wolf waiting for him. As he neared her, he became suddenly cautious. He slowed down to an alert and mincing walk and then stopped. He regarded her carefully and dubiously, yet desirefully. She seemed to smile at him, showing her teeth in an ingratiating rather than a menacing way. She moved toward him a few steps, playfully, and then halted. One Ear drew near to her, still alert and cautious, his tail and ears in the air, his head held high.

He tried to sniff noses with her, but she retreated playfully and coyly. Every advance on his part was accompanied by a corresponding retreat on her part. Step by step she was luring him away from the security of his human companionship. Once, as though a warning had in vague ways flitted through his intelligence, he turned his head and looked back at the overturned sled, at his team-mates, and at the two men who were calling to him.

But whatever idea was forming in his mind, was dissipated by the she-wolf, who advanced upon him, sniffed noses with him for a fleeting instant, and then resumed her coy retreat before his renewed advances.

In the meantime, Bill had bethought himself of the rifle. But it was jammed beneath the overturned sled, and by the time Henry had helped him to right the load, One Ear and the she-wolf were too close together and the distance too great to risk a shot.

Too late One Ear learned his mistake. Before they saw the cause, the two men saw him turn and start to run back toward them. Then, approaching at right angles to the trail and cutting off his retreat they saw a dozen wolves, lean and grey, bounding across the snow. On the instant, the she-wolf's coyness and playfulness disappeared. With a snarl she sprang upon One Ear. He thrust her off with his shoulder, and, his retreat cut off and still intent on regaining the sled, he altered his course in an attempt to circle around to it. More wolves were appearing every moment and joining in the chase. The she-wolf was one leap behind One Ear and holding her own.

"Where are you goin'?" Henry suddenly demanded, laying his hand on his partner's arm.

Bill shook it off. "I won't stand it," he said. "They ain't a-goin' to get any more of our dogs if I can help it."

Gun in hand, he plunged into the underbrush that lined the side of the trail. His intention was apparent enough. Taking the sled as the centre of the circle that One Ear was making, Bill planned to tap that circle at a point in advance of the pursuit. With his rifle, in the broad daylight, it might be possible for him to awe the wolves and save the dog.

"Say, Bill!" Henry called after him. "Be careful! Don't take no chances!"

GO ON TO THE NEXT PAGE

Henry sat down on the sled and watched. There was nothing else for him to do. Bill had already gone from sight; but now and again, appearing and disappearing amongst the underbrush and the scattered clumps of spruce, could be seen One Ear. Henry judged his case to be hopeless. The dog was thoroughly alive to its danger, but it was running on the outer circle while the wolf-pack was running on the inner and shorter circle. It was vain to think of One Ear so outdistancing his pursuers as to be able to cut across their circle in advance of them and to regain the sled.

The different lines were rapidly approaching a point. Somewhere out there in the snow, screened from his sight by trees and thickets, Henry knew that the wolf-pack, One Ear, and Bill were coming together. All too quickly, far more quickly than he had expected, it happened. He heard a shot, then two shots, in rapid succession, and he knew that Bill's ammunition was gone.

Questions 23–30

23 Which of the following BEST describes Henry's feelings in the story?

    A   anxious, then hopeful

    B   curious, then frightened

    C   worried, then discouraged

    D   calm, then outraged

24 This story is MOSTLY about

    A   friends who are training for a dog sled race in the winter

    B   two men and a team of dogs that encounter a pack of wolves

    C   a trained sled dog that wants to escape into the wilderness

    D   dog sled racers who accidentally overturn their sled

GO ON TO THE NEXT PAGE

25 Why did One Ear sneak away from Bill and Henry?

  A   He caught the scent of meat and followed it.

  B   He was tired and wanted to go to sleep.

  C   He started following another sled's trail and got lost.

  D   He was lured away by a female wolf.

26 Bill's and Henry's MAIN conflict in this story is that

  A   they have overturned their sled in the forest.

  B   their dogs are too tired to pull their sled fast.

  C   they are being stalked by a pack of hungry wolves.

  D   their dogs have pulled the sled off the main trail.

27 When the author says that Bill "seemed to have forgotten his forebodings of the previous night," he means that Bill

  A   isn't worried that they are moving slower than planned.

  B   doesn't care that some of the dogs have been killed.

  C   isn't concerned about the wolves anymore.

  D   doesn't remember what Henry said to him.

28 The mood of this story is

  A   suspenseful.

  B   peaceful.

  C   depressing.

  D   frustrating.

GO ON TO THE NEXT PAGE

29 What is the theme of the story?

    A Friends should take care of one another.

    B Things are not always as they seem.

    C Fighting is not the way to solve problems.

    D Man and nature are in constant struggle.

30 In the excerpt from *White Fang*, Jack London compares

    A dogs and people.

    B friends and enemies.

    C daylight and darkness.

    D sharks and wolves.

GO ON TO THE NEXT PAGE

DIRECTIONS: Read Selection 2, "How to Set Up an Aquarium." Then answer the questions that follow.

How to Set Up an Aquarium

Have you ever owned a pet fish? Compared to most popular domestic animals, fish are low-maintenance creatures. They're well-behaved, too. They won't gnaw on furniture, shred curtains, or bark at the neighbors!

Setting up an aquarium can be an enjoyable project that calls on you not only to choose the conditions that would most benefit the fish, but also to make creative decisions that turn an aquarium into a piece of aquatic art. A decorative aquarium can improve the appearance of your home with just a little bit of effort.

In order to construct an aquarium that's safe for fish and pleasing to the eye, follow these general guidelines. For more specific information, consult a specialist at your local pet shop.

Getting Started

You'll need a number of materials in order to get started. The most important item is, of course, the aquarium itself. Aquariums come in all shapes and sizes, from traditional one-gallon fishbowls meant for a single swimmer to massive tanks that can hold dozens of fish. In order to select an appropriately sized aquarium, consider how many fish you intend to keep in it. To allow your fish to live comfortably, provide at least one gallon of water per fish.

What You'll Need

 After you've chosen the best aquarium, you'll need some special equipment to make it a safe and healthy habitat for your fish. You'll need aquarium gravel, a water filter, a water heater, a floating thermometer, and a pump.

GO ON TO THE NEXT PAGE

Step by Step

Follow these step-by-step instructions to set up a beautiful aquarium in your home.

Step 1: Once you've acquired the necessary materials, the first step is to cleanse the aquarium of any grime, sediments, or other refuse that may have accumulated in it. Avoid using cleaning chemicals, though, since they can contaminate the water you later add to the aquarium.

Step 2: Once the aquarium is clean, add gravel to the bottom, typically one pound per gallon of water. You can even accessorize your aquarium with rocks, plants, or fanciful ornaments.

Step 3: You'll want to install a filter in order to remove contaminants from the water and keep your fish healthy. Select a filter that's suitable for the size of your aquarium, and then install it according to the directions.

Step 4: The next step is to fill the aquarium with clean, cool water; a safe guideline here is to only utilize water that you would consider drinkable. Don't fill the aquarium right to the top, though, because there are still a few more items you'll need to add, including the water heater and pump.

Step 5: Install your water heater and pump according to the directions on the packages of these appliances. Usually, the heater should be adjusted to keep the water at a temperature of about seventy-five degrees Fahrenheit.

Step 6: Now your fish should be comfortable and healthy—unless you forget to add them! The most crucial component of an aquarium is, of course, the fish. Add them to the water and enjoy your new flippered friends.

Questions 31–38

31 The steps in the article are useful because they

A make all of the information in the article look the same.

B separate information into smaller sections that are easy to understand.

C save space for the graphics and pictures that explain the steps.

D make the information in the article more interesting for readers.

GO ON TO THE NEXT PAGE

32 How is the information in this article organized?

    A   comparison and contrast

    B   cause and effect

    C   sequence

    D   order of importance

33 The article says to clean the aquarium before adding materials to it. You should NOT use cleaning chemicals to do this because they

    A   will not remove all of the grime from the sides of the aquarium.

    B   are not made for cleaning aquariums.

    C   might get into the water and affect the fish.

    D   will make the sides of the glass look cloudy.

34 According to the article, why does an aquarium need a floating thermometer?

    A   to make sure that the water stays at seventy-five degrees

    B   to make sure the temperatures of the water and outside air are the same

    C   to create an ocean environment for the fish

    D   to decorate the aquarium with something other than plants

35 This article was probably written by a

    A   pet store employee.

    B   marine scientist.

    C   veterinarian.

    D   goldfish owner.

36   The following sentence appears in the fifth paragraph of the article.

After you've chosen the best aquarium, you'll need some special equipment to make it a safe and healthy habitat for your fish.

Based on how the word is used in the article, which of the following BEST describes the meaning of "habitat"?

A   environment

B   location

C   temperature

D   lifespan

37   Which idea from the text does the photograph of the fish tank support?

A   You need a filter in your tank to keep your fish healthy.

B   You should generally provide at least one gallon of water per fish.

C   Grime and sediment should be cleaned from the tank before the fish are added.

D   A decorative aquarium can improve the appearance of your home.

38   The author begins the passage with a question to

A   show that fish are friendly.

B   get the reader's attention.

C   introduce an explanation.

D   provide background information.

STOP

PART 3—READING

LINKED SELECTION

DIRECTIONS: Read this selection and answer the questions that follow. **You will have 30–50 minutes to complete this section.**

The Dragons of Ancient China

Throughout history, people all around the world have been fascinated with dragons. There have been thousands of narratives based on the larger-than-life flying lizards and the lion-hearted heroes who interact with them. People across the globe find dragons captivating and compelling, but nowhere have dragons ever been as celebrated as in ancient China.

The ancient Chinese perspective regarding dragons is one you may not expect, however. In America, dragons are typically portrayed as menacing and villainous monsters who crush villages, trample castles, and spew fiery breath at any heroes who dare to challenge them. In ancient China, however, the dragons *were* the heroes. These mythical dragons represented every positive characteristic people admired. They were wise, strong, compassionate, and beautiful. In many ways, the Lung, or dragon, is a symbol of the nation of China.

The ancient Chinese, from peasants to royalty, believed that the Lungs protected their lands and families and assisted them at all times. In fact, the Chinese believed that the people of their nation were descendants of dragons. They believed that Lungs actually created the

GO ON TO THE NEXT PAGE

Chinese people. The ruling classes and royalty in particular felt a connection to the Lungs. Emperors throughout Asia have claimed to have dragons in their families. For instance, Emperor Hirohito, the leader of Japan from 1926 to 1989, believed that the Dragon King of the Sea created his family 2,500 years ago.

According to myth, dragons not only established the royal families of Asia but also continued to advise them. In the 1200s, a king of Cambodia spent hours at a time locked in a golden tower, supposedly conversing with a nine-headed dragon. The greatest emperors were even thought to be dragons themselves! For centuries, people in Japan were not allowed to observe their emperors. People believed this was because the emperors had transformed into *Lungs* too glorious to look upon.

Today in America, if someone called you "dragon face," you would probably be highly insulted because the phrase suggests of ugliness or meanness. If you lived in ancient China, however, if someone called you "dragon face," you would consider it a great compliment. The phrase "dragon face" would suggest that you were extraordinarily beautiful like a dragon. In fact, ancient China anything compared to a dragon was considered beautiful. For example, dragon-house, dragon-throne, and dragon-land would be considered great compliments. According to ancient Chinese tradition, there is even a Year of the Dragon during which great prosperity will come to people, especially those born in that year.

Dragons could be seen everywhere in ancient China. They were carved onto musical instruments because people believed their loved music. They were inscribed on books and tablets, because people believed they had a flair for literature. They were even carved outside of temples because people believed they would protect holy places. Thrones, bridges, and swords were all decorated with dragons. The people of ancient Asia wanted dragons to be involved in all aspects of their lives.

These legendary lizards had appearances just as varied as their tasks. They could have skins of any color of the rainbow, though usually they were green or golden. Some sported horns, wings, gigantic teeth, or even catlike whiskers which would help the dragon move around in the deep, dark oceans. For the most part, dragons' body parts resembled the parts of other animals. Many dragons shared characteristics of animals like bulls, frogs, tigers, eagles, and even camels and rabbits. Some dragons began their lives as carp, a type of fish. Other dragons created unique breeds of animals by mating with the creatures of the earth.

People believed that dragons loved to assist them, but only if people appreciated them. If people were unappreciative, a dragon might cause a flood or a drought to punish them. Because of this, the people of ancient Asia went to great lengths to honor dragons. They even had parades during which people would wear elaborate dragon disguises to celebrate the powerful Lungs. These parades are still held today, which shows the lasting influence of this fascinating Chinese tradition.

GO ON TO THE NEXT PAGE

Questions 39–46

39 Which sentence from the article BEST supports the idea that dragons "demanded that the people showed their appreciation"?

A   "The dragon might cause floods or droughts to punish them."

B   "They believed that Lungs actually created the Chinese people."

C   "These mythical dragons represented every positive characteristic the people admired."

D   "One dragon was thought to help people manage their money; another brought literacy to the peasants."

40 The following sentence appears in the sixth paragraph of the article.

They were inscribed on books and tablets because people believed they had a flair for literature.

Based on how the phrase is used in the article, which of the following BEST describes the meaning of *flair for literature*?

A   great at writing

B   love of reading

C   looked good on

D   spark of intelligence

41 From the author's point of view, the worldwide appeal of dragon legends is especially interesting because

A   most legends focus on human characters.

B   people always realized dragons were imaginary.

C   the dragons mean different things to different people.

D   thousands of movies have been made about dragons.

GO ON TO THE NEXT PAGE

42 What did the body parts of dragons in ancient China usually resemble?

A large lizards

B other animals

C human beings

D a type of fish

43 If someone in ancient China referred to a city as a "dragon-land," this would probably mean the city was

A scary.

B very old.

C very large.

D beautiful.

44 Which is the BEST heading for the section containing paragraphs 2 and 3?

A "Emperor Hirohito"

B "Dragons in America"

C "Dragons in Ancient China"

D "The Dragon King of the Sea"

45 Why did the author write this article?

A to describe different types of dragons

B to compare and contrast views about dragons

C to entertain readers with a story about dragons

D to inform readers about dragons in ancient China

GO ON TO THE NEXT PAGE

46 People in Japan believed they were NOT allowed to see their emperors because

    A   they had turned into dragons.

    B   they were reading about dragons.

    C   they were spending time with dragons.

    D   they were drawing pictures of dragons.

STOP

PART 4—WRITING

PART 4A: WRITING FROM KNOWLEDGE AND EXPERIENCE

DIRECTIONS:

In Part 4A, you will be given a theme and a number of ways to write about it. You must choose ONLY ONE way.

We will begin Part 4A together by reading the information on the next page. As I read aloud, please follow along on the page.

WAIT. DO NOT GO ON UNTIL TOLD TO DO SO.

Writing 4A: WRITING FROM KNOWLEDGE AND EXPERIENCE

---

47   BEING A HERO

Most of us know someone whom we consider a hero. This person may have done something great or may do small things to help people each day. This person might be a teacher, parent, or good friend. Write about helping others and being a hero.

Do **ONLY ONE** of the following:

describe a situation in which someone helped you

OR

define what it means to be a hero

OR

explain why you think it is important to help others

OR

persuade the reader that someone you know is a hero

OR

write about the theme in your own way.

You may use examples from real life, from what you read or watch, or from your imagination.

Your audience will be interested adults.

You may use pages 35–36 for writing down ideas, organizing your thoughts, or writing a rough draft. Use the checklist on page 34 to help you improve your writing. Pages 64 and 65 contain the scorepoint descriptions used by readers to score your writing.

The final copy of your response must be written in the lined spaces starting on page 50. Only the writing on this page will be scored. No additional sheets may be used. Nothing written on the prewriting and rough draft pages will be scored.

GO ON TO THE NEXT PAGE

CHECKLIST FOR REVISING AND PROOFREADING

DIRECTIONS:

Use the following checklists as you revise and proofread the writing you have done for Part 4A. When you are finished revising, you must write your final copy. Then, proofread your final copy to make sure that all of your revisions have been made.

CHECKLIST FOR REVISION:

_____ Do I have a clear central idea that connects to the theme?

_____ Do I stay focused on the theme?

_____ Do I support my central idea with important details/examples?

_____ Do I need to take out details/examples that DO NOT support my central idea?

_____ Is my writing organized and complete?

_____ Do I use a variety of words, phrases, and/or sentences?

CHECKLIST FOR EDITING:

_____ Have I checked and corrected my spelling to help readers understand my writing?

_____ Have I checked and corrected my punctuation and capitalization to help readers understand my writing?

CHECKLIST FOR PROOFREADING:

_____ Is everything in my final copy the way that I want it?

GO ON TO THE NEXT PAGE

NOTES/PLANNING

GO ON TO THE NEXT PAGE

NOTES/PLANNING

PART 4B: STUDENT WRITING SAMPLE

DIRECTIONS:

In Part 4B, you will read a student writing sample, answer the multiple-choice questions that follow, and then write a short response. You may look back at the student writing sample as often as needed during Part 4B.

You may underline, highlight, or write notes in this booklet, but you must mark all of your answers in Part 4B of your **Answer Sheets** on page 52.

You may not use any resource materials (dictionaries, grammar books, spelling books, etc.) for any part of this test.

When you have finished Part 4B, STOP.

WAIT. DO NOT GO ON UNTIL TOLD TO DO SO.

DIRECTIONS: Read the following passage and answer the questions that follow. **You will have about 30 minutes to complete this section.**

"The Lake"

(1)　　　　I am writing about a favorite place to visit. This story is about my Uncle
(2)　Ray's cottage by a lake. In the summer when the whether is warm, my family
(3)　and I spend a weekend or two at Uncle Ray's. The lake is really beautiful in the
(4)　summer. It is large and surounded by tall trees. In the morning, when the sun
(5)　rises, it makes the lake sparkle. You can hear all kinds of birds singing as they
(6)　welcome the new day.
(7)　　　　There is a lot to do at the lake. My Uncle Ray has a small boat and lots
(8)　of fishing poles. We go out in the middle of the lake and fish for hours. It is
(9)　really fun. There is also a dock near his yard where you can jump off in. The
(10)　water in the lake is cool even in the summer it feels nice on a hot summer day.
　　　My sister likes to kayak around the lake. She likes to watch the wildlife.

Questions 48–52

48 Choose another title for this piece.

A "Great Fishing"

B "Pretty Mornings"

C "Swimming in the Lake"

D "A Fun Getaway"

49 What type of genre is reflected in this piece?

A Short story

B Memoir

C Biography

D Myth

GO ON TO THE NEXT PAGE

50 In which lines does the author use adjectives to describe the lake?

   A   Lines 1 and 2

   B   Lines 3 and 4

   C   Lines 5 and 6

   D   Lines 7 and 8

51 What is the correct spelling of the word <u>whether</u> (Line 2)?

   A   whethering

   B   weather

   C   wether

   D   wither

52 What is the correct spelling of <u>surounded</u> (Line 4)?

   A   suronded

   B   surrounded

   C   surroundded

   D   serounded

GO ON TO THE NEXT PAGE

PEER RESPONSE TO THE STUDENT WRITING SAMPLE

DIRECTIONS: Write a response to the question in the box below. You may look back at the student writing sample as often as needed during Part 4B.

---

53 **Where do you think the author of this essay lives? How is it different from where Uncle Ray lives?**

Use details from the student writing sample to support your answer.

Use the checklist on the next page to help you with your response. The Notes/Planning space may be used for writing down and organizing your ideas. Your response must be written in the lined spaces starting on page 53 of your **Answer Sheets**. Only the writing on your **Answer Sheets** will be scored. No extra sheets may be used.

You may not use any resource materials (dictionaries, grammar books, spelling books, etc.) for any part of this test.

When you have finished Part 4B, STOP.

I would guess that the author probably lived in an urban area. The author's Uncle probably lived in a tropical area, and the tall trees the author mentioned could possibly be palm trees. The author could have possibly gone on vacation to a tropical place, because it was winter where he/her lived.

PART 4B: CHECKLIST FOR THE
PEER RESPONSE TO THE STUDENT WRITING SAMPLE

DIRECTIONS:

Use this checklist to help you with your response. Your response must be written in the lined spaces starting on page 53 of your **Answer Sheets**.

CHECKLIST:

_____   Do I clearly answer the question that was asked?

_____   Do I support my answer with details from the student writing sample?

_____   Is my response complete?

GO ON TO THE NEXT PAGE

NOTES/PLANNING

NOTES/PLANNING

PART 1A: PAIRED READING SELECTIONS

MARKING INSTRUCTIONS

Make heavy BLACK marks.
Erase cleanly.
Make no stray marks.

● CORRECT MARK

◉ ⊘ ⊗ ◖ INCORRECT MARK

Multiple-choice questions

1. (A) (B) (C) (D)
2. (A) (B) (C) (D)
3. (A) (B) (C) (D)
4. (A) (B) (C) (D)
5. (A) (B) (C) (D)
6. (A) (B) (C) (D)
7. (A) (B) (C) (D)
8. (A) (B) (C) (D)
9. (A) (B) (C) (D)
10. (A) (B) (C) (D)
11. (A) (B) (C) (D)

12. (A) (B) (C) (D)
13. (A) (B) (C) (D)
14. (A) (B) (C) (D)
15. (A) (B) (C) (D)
16. (A) (B) (C) (D)
17. (A) (B) (C) (D)
18. (A) (B) (C) (D)
19. (A) (B) (C) (D)
20. (A) (B) (C) (D)
21. (A) (B) (C) (D)

Student Name_____

PART 1B: RESPONSE TO THE PAIRED READING SELECTIONS

Write your final response for question 22 here.

Write your final response for question 22 here.

PART 2: INDEPENDENT READING SELECTIONS

MARKING INSTRUCTIONS

Make heavy BLACK marks.
Erase cleanly.
Make no stray marks.

●   ◉   ⊘   ⊗   ◖
CORRECT   INCORRECT
MARK   MARK

Multiple-choice questions

23. Ⓐ   Ⓑ   Ⓒ   Ⓓ   31. Ⓐ   Ⓑ   Ⓒ   Ⓓ

24. Ⓐ   Ⓑ   Ⓒ   Ⓓ   32. Ⓐ   Ⓑ   Ⓒ   Ⓓ

25. Ⓐ   Ⓑ   Ⓒ   Ⓓ   33. Ⓐ   Ⓑ   Ⓒ   Ⓓ

26. Ⓐ   Ⓑ   Ⓒ   Ⓓ   34. Ⓐ   Ⓑ   Ⓒ   Ⓓ

27. Ⓐ   Ⓑ   Ⓒ   Ⓓ   35. Ⓐ   Ⓑ   Ⓒ   Ⓓ

28. Ⓐ   Ⓑ   Ⓒ   Ⓓ   36. Ⓐ   Ⓑ   Ⓒ   Ⓓ

29. Ⓐ   Ⓑ   Ⓒ   Ⓓ   37. Ⓐ   Ⓑ   Ⓒ   Ⓓ

30. Ⓐ   Ⓑ   Ⓒ   Ⓓ   38. Ⓐ   Ⓑ   Ⓒ   Ⓓ

Student Name _____

PART 3: LINKED SELECTION

MARKING INSTRUCTIONS

Make heavy BLACK marks.
Erase cleanly.
Make no stray marks.

CORRECT
MARK

INCORRECT
MARK

Multiple-choice questions

39. Ⓐ Ⓑ Ⓒ Ⓓ   43. Ⓐ Ⓑ Ⓒ Ⓓ

40. Ⓐ Ⓑ Ⓒ Ⓓ   44. Ⓐ Ⓑ Ⓒ Ⓓ

41. Ⓐ Ⓑ Ⓒ Ⓓ   45. Ⓐ Ⓑ Ⓒ Ⓓ

42. Ⓐ Ⓑ Ⓒ Ⓓ   46. Ⓐ Ⓑ Ⓒ Ⓓ

Student Name_____

PART 4A: WRITING FROM KNOWLEDGE AND EXPERIENCE

Write your final response for question 47 here.

Write your final response for question 47 here.

PART 4B: STUDENT WRITING SAMPLE

MARKING INSTRUCTIONS

Make heavy BLACK marks.
Erase cleanly.
Make no stray marks.

●   ◉   ⊘   ⊗   ◓
CORRECT   INCORRECT
MARK   MARK

Multiple-choice questions

48. Ⓐ   Ⓑ   Ⓒ   Ⓓ

49. Ⓐ   Ⓑ   Ⓒ   Ⓓ

50. Ⓐ   Ⓑ   Ⓒ   Ⓓ

51. Ⓐ   Ⓑ   Ⓒ   Ⓓ

52. Ⓐ   Ⓑ   Ⓒ   Ⓓ

Student Name_____

Write your final response for question 53 here.

Write your final response for question 53 here.

ANSWER KEY

Pretest Answers

PART 1—READING

Part 1A: Paired Reading Selections

1   **D**   **R.NT.08.03**   **Narrative Text**
Meg complains about many small things in the excerpt, but the primary reason she is unhappy is discussed in the first paragraph. She wishes she could spend more time having fun.

2   **B**   **R.CS.08.01**   **Critical Standards**
The author of the passage would most likely agree that girls who have no work to do are very lucky. The narrator tells the story of the girls objectively, meaning the author doesn't reveal her opinions about the characters' actions to the reader. The only idea that the author seems to agree with is answer choice B.

3   **A**   **R.CM.08.03**   **Comprehension**
This excerpt contains many ideas, but the main idea is that having people we love gets us through hard times. In the excerpt, some of the other points are made, but they are not the main idea.

4   **C**   **R.CM.08.01**   **Comprehension**
In the excerpt, the characters run around and attempt to get ready to leave the house. None of the girls seem to have enough time to do what they want to do. This action causes the girls to be confused and upset. Preparing to leave is very busy and hectic.

5   **B**   **R.CM.08.02**   **Comprehension**
Answer choices C and D are incorrect because they are untrue. Meg is not laughing with Jo and never says that she cannot find the ribbon. Meg does not put a ribbon around her neck because she feels she no one to look pretty for. Answer choice B is correct.

6   **D**   **R.NT.08.03**   **Narrative Text**
In the beginning of the excerpt, Meg points out that she wishes she did not have to work, but could have fun like other girls do. She complains that she has to work so hard and decides not to look nice. It is her mother who asks the girls to be quiet. Answer choice D is correct.

7   A   R.NT.08.04   Narrative Text
Answer choice B is incorrect. Marmee only asks the girls to be quiet once in the excerpt. Though answer choices C and D may be correct, there is no evidence for those choices in this excerpt. In paragraph 2, Jo suggests the girls "shoulder our bundles . . . as cheerfully as Marmee does," meaning that they should live by their mother's example.

8   B   R.CM.08.02 Comprehension
All of the choices are somehow alluded to in the excerpt. But the paragraphs serve as a turning point in the excerpt. All of the girls are upset until Hannah brings out the turnovers, which shows that small things can sometimes cheer people. Answer choice B is correct.

9   A   R.CM.08.01 Comprehension
Answer choices C and D are not true at all. The speaker is not sad and lonely; she is writing about her love. The speaker is not angry and sullen, but happy to be expressing her love. Though she is happy, she does not seem selfish since she wouldn't trade love for possessions. The speaker must be loving, since she'd forgo material things for love. She is also clever in leading readers toward one ending and then surprising them with another.

10   B   R.NT.08.02   Narrative Text
In the poem, the speaker does not contrast love and hate or friendship and romance. The idea of death comes up, not in contrast to love, but as a result of the lack of love. In the poem, the speaker compares love to other necessities of life.

11   C   R.NT.08.04   Narrative Text
This poem does not have a disappointed mood because the speaker has not been let down in any way. The poem does not represent any resentful feelings either. Although the speaker is serious about her love, her mood is not serious. The speaker is rather playful in her method of expression, tricking her readers in the last line.

12   D   R.CM.08.03 Comprehension
The poet is making the point that if a man were sinking, love would not keep him afloat. Therefore, when the poet says that love is not "a floating spar to men that sink," she is contrasting love to something that helps men stay above water. Answer choice D is correct.

13   D   R.NT.08.03   Narrative Text
Though this poem does list the hardships that might overcome our need for love for a short time, the overall message is that love is actually more important than all of those things. D is the correct answer.

14   B   R.NT.08.01   Narrative Text
Answer choices A, C, and D are incorrect. Throughout the poem, the speaker is making the point that love cannot prevent pain or hunger, but it is extremely important anyway. B is the correct answer.

15   **C**   **R.NT.08.04**   **Narrative Text**
Answer choices B and D are incorrect because they both illustrate the point that love is the most important thing to the author. Answer choice A is not specifically about the author, but just a statement about love in general. Only answer choice C suggests that love might not be the most important thing for the author. She might trade it for peace.

16   **C**   **R.CM.08.02**   **Comprehension**
In the last few lines of the poem, the author lists a few extreme things that might be more important to her than love but, in the final line, insists that love is the most important. She implies that even if she were desperate, she would not trade a memory for food, because a memory is dearer to her than food. The correct answer is C.

17   **C**   **R.CM.08.03**   **Comprehension**
The speaker of the poem says that she would not trade her happy memory for anything tangible or materialistic, and in the excerpt from *Little Women*, even though the girls are very poor, they cherish the one last look at their mother each morning. Answer choice C is the best answer.

18   **A**   **R.CM.08.03**   **Comprehension**
Both passages include a description of someone who is hurting. The speaker of the poem says that men may be dying for lack of love. In the excerpt from *Little Women*, Meg is hurting because she must spend so much time with the younger children.

19   **D**   **R.CM.08.03**   **Comprehension**
While answer choice A may seem correct, we don't know for certain that the speaker of the poem is rich. We do know, however, that she is mature, meaning she is an adult. Answer choice D is the best answer.

20   **A**   **R.CM.08.03**   **Comprehension**
The speaker of the poem uses "I," so the poem is in the first person. In the excerpt from *Little Women*, the narrator is someone other than the characters in the story, so it is written in the third person. Answer choice A is the correct answer.

21   **B**   **R.CM.08.03**   **Comprehension**
You can determine the correct answer to this question using process of elimination. The only idea explored in both passages is people wanting more than they have: answer choice B.

Part 1B: Response to the Paired Reading Selections

W.PR. 08.01, .02, .03, .04, .05	Writing Process
W.GR.08.01	Grammar and Usage
W.SP.08.01	Spelling
W.HW.08.01	Handwriting

Michigan Educational Assessment Program (MEAP)
Grades 3–8

Rubric for the Response to the Paired Reading Selections

6	The student clearly and effectively chooses key or important ideas from each reading selection to support a position on the question and to make a clear connection between the reading selections. The position and connection are thoroughly developed with appropriate examples and details. There are no misconceptions about the reading selections. There are strong relationships among ideas. Mastery of language use and writing conventions contributes to the effect of the response.
5	The student makes meaningful use of key ideas from each reading selection to support a position on the question and to make a clear connection between the reading selections. The position and connection are well developed with appropriate examples and details. Minor misconceptions may be present. Relationships among ideas are clear. The language is controlled, and occasional lapses in writing conventions are hardly noticeable.
4	The student makes adequate use of ideas from each reading selection to support a position on the question and to make a connection between the reading selections. The position and connection are supported by examples and details. Minor misconceptions may be present. Language use is correct. Lapses in writing conventions are not distracting.
3	The student takes a clear position on the question. The response makes adequate use of ideas from one reading selection **or** partially successful use of ideas from both reading selections to support the position. The position is developed with limited use of examples and details. Misconceptions may indicate only a partial understanding of the reading. Language use is correct but limited. Incomplete mastery over writing conventions may interfere with meaning some of the time.
2	The student takes a clear position on the question. There is partially successful use of ideas from one reading selection **or** minimal use of ideas from both reading selections to support the position. The position is underdeveloped. Major misconceptions may indicate minimal understanding of the reading. Limited mastery over writing conventions may make the writing difficult to understand.

1	The student takes a position on the question but only makes minimal use of ideas from one reading selection **or** the student attempts to support an unclear position with minimal use of ideas from both reading selections. Ideas are not developed and may be unclear. Major misconceptions may indicate a lack of understanding of the reading. Lack of mastery over writing conventions may make the writing difficult to understand.

22   The Definition of Love

Sample 6-point answer: Love is a feeling of affection people have for one another. When people love others and are loved in return, they feel secure and happy. In the excerpt from *Little Women*, the sisters are quarreling. They dread having to go outside on a long walk with Mrs. March. Hannah makes turnovers each time they go, because she knows the walk will be long and cold. However, at the end of the excerpt, the narrator says that the girls always look back before turning the corner to see their mother at the window and watch her nod and smile. The narrator says that it seemed they couldn't get through the day without that and that last glimpse of that motherly face affected them like sunshine. The girls love their mother, and knowing that she loves them in return makes them feel good.

The speaker in "Love Is Not All" discusses what love is not: it is not a house or air, or food. She contemplates whether, in a difficult time, she would trade someone's love or a memory of time spent with this person for something she needs to stay alive. She concludes that she would not. She shows that while love isn't a tangible necessity, it is still a necessity.

PART 2—READING

Part 2: Independent Reading Selections

23   C   R.NT.08.01   Narrative Text
In the beginning of the story, both Bill and Henry seem worried about the wolves following their sled. Later, Henry watches Bill run after the wolves and feels hopeless about their situation. Answer choice C is the best answer.

24   B   R.CM.08.02 Comprehension
The story never mentions that Bill and Henry are in a race, which eliminates choices A and D. While One Ear does get distracted by the she-wolf, he's not looking to escape into the wild. The correct answer is B. Bill, Henry, and their dogs are being pursued by a pack of wolves.

25   D   R.CM.08.02 Comprehension
One Ear got distracted by the she-wolf. She was acting playful and led him away from the safety of his owners.

26   C   R.NT.08.03   Narrative Text

Answer choice D is not a conflict in the story. Choices A, B, and C are all conflicts in the story. However, the main conflict is that Bill and Henry are being followed by a pack of wolves that keeps attacking their dogs.

27   C   R.WS.08.01 Word Study

In this sentence, the word "forebodings" means concerns that something bad might happen. The previous night, Bill was worried that he and Henry would lose another dog to the wolves. However, when he wakes up, he seems happier and less worried. The correct answer is C.

28   A   R.NT.08.04   Narrative Text

The mood of this story is certainly not peaceful. Bill and Henry are being followed by a pack of wolves that keeps attacking their dogs. While it may seem depressing or frustrating in some parts, the cliffhanger ending makes the mood in the story suspenseful.

29   D   R.NT.08.03   Narrative Text

This story illustrates the struggle between man and nature. Man is represented by Bill and Henry. Nature is represented by the wolves in the wilderness.

30   D   R.IT.08.02   Informational Text

In the beginning of the story, Bill remarks that he's heard sailors talk about sharks following their ships. He calls the wolves "land sharks" because they've been following their sled.

31   B   R.IT.08.01   Informational Text

Organizing information into steps helps readers understand smaller portions of information at a time. Answer choice B is correct.

32   C   R.IT.08.02   Informational Text

Steps tell readers the order in which each piece of a project is supposed to occur. This is a sequence method.

33   C   R.CM.08.01 Comprehension

Step 1 of the article advises readers not to use cleaning chemicals to clean the aquarium because the chemicals can contaminate the water. Students can infer that contaminated water would affect the fish.

34   A   R.IT.08.02   Informational Text

Step 5 of the article says that the water should be kept at a temperature of about seventy-five degrees. Answer choice A is correct.

35   C   R.IT.08.01   Informational Text

The article contains specific information that a pet store employee or goldfish owner might not know, such as the ideal temperature of the water. A marine scientist would not write about keeping fish as pets. A veterinarian would know the information in the article and might write about keeping fish as pets.

36 **A** **R.WS.08.01 Word Recognition and Word Study**
The article talks about creating a good environment for fish. Though it does discuss some of the other elements mentioned in the answer choices, students should be able to infer that the word *habitat* refers to the fish's environment.

37 **D** **R.IT.08.03 Informational Text**
The photograph illustrates the decorative aspect of an aquarium. It does not focus on a filter, the fish-to-water ratio of the tank, or cleaning the tank.

38 **B** **R.IT.08.03 Informational Text**
When an author asks the reader a question at the beginning of an article, the reader tends to pay more attention from the beginning of the article. This technique tends to draw readers into the text on a more personal level.

PART 3—READING

Part 3: Linked Selection

39 **A** **R.CS.08.01 Critical Standards**
Answer choice A best supports the idea that dragons "demanded that people showed their appreciation" because it states that dragons might punish people with floods or droughts.

40 **B** **R.WS.08.07 Word Study**
People inscribed dragons on books and tablets because they thought the dragons might want to read these items, so a "flair for literature" means a love of reading.

41 **C** **R.CM.08.03 Comprehension**
The author seems interested in dragons because they mean different things to different people. Answer choice C is correct.

42 **B** **R.CM.08.03 Comprehension**
The passage says that the body parts of dragons in ancient China usually resembled other animals, so answer choice B is correct.

43 **D** **R.WS.08.07 Word Study**
The word "dragon" in ancient China meant "beauty" and the passage said if someone called you "dragon-face" it would be a compliment. Therefore, a "dragon-land" would be a beautiful land.

44 **C** **R.CM.08.02 Comprehension**
Paragraphs 2 and 3 discuss dragons in Ancient China, so answer choice C is the correct answer.

45 **D** **R.CS.08.01 Critical Standards**
The passage is informational and it gives readers information about dragons in ancient China.

46 **A** **R.CM.08.02 Comprehension**
This answer is stated in the passage. When people in ancient Japan were not allowed to see their emperors, they believed it was because their emperors had turned into dragons.

PART 4—WRITING

Part 4A: Writing from Knowledge and Experience

Michigan Educational Assessment Program (MEAP)

W.PR. 08.01, .02, .03, .04, .05	Writing Process
W.GR.08.01	Grammar and Usage
W.SP.08.01	Spelling
W.HW.08.01	Handwriting

Writing from Knowledge and Experience
Grade 8

Holistic Scorepoint Descriptions

6	The writing is exceptionally clear and focused. Ideas and content are thoroughly developed with relevant details and examples where appropriate. The writer's control over organization and the connections between ideas moves the reader smoothly and naturally through the text. The writer shows a mature command of language including precise word choice that results in a compelling piece of writing. Tight control over language use and mastery of writing conventions contribute to the effect of the response.
5	The writing is clear and focused. Ideas and content are well developed with relevant details and examples where appropriate. The writer's control over organization and the connections between ideas effectively moves the reader through the text. The writer shows a command of language including precise word choice. The language is well controlled, and occasional lapses in writing conventions are hardly noticeable.
4	The writing is generally clear and focused. Ideas and content are developed with relevant details and examples where appropriate, although there may be some unevenness. The response is generally coherent, and its organization is functional. The writer's command of language, including word choice, supports meaning. Lapses in writing conventions are not distracting.

3	The writing is somewhat clear and focused. Ideas and content are developed with limited or partially successful use of examples and details. There may be evidence of an organizational structure, but it may be artificial or ineffective. Incomplete mastery over writing conventions and language use may interfere with meaning some of the time. Vocabulary may be basic.
2	The writing is only occasionally clear and focused. Ideas and content are underdeveloped. There may be little evidence of organizational structure. Vocabulary may be limited. Limited control over writing conventions may make the writing difficult to understand.
1	The writing is generally unclear and unfocused. Ideas and content are not developed or connected. There may be no noticeable organizational structure. Lack of control over writing conventions may make the writing difficult to understand.

47   My Mother Is a Hero

Sample 6-point response: My mother is definitely a hero in my eyes. She is a single mother and she does her best to be there for me and support us. My mother had me before she finished college, but she believes that education is very important, so she takes classes whenever she has time. Her goal is to one day finish her degree. She has also instilled the value of education in me. I plan to study hard and go to college one day myself.

My mother has also taught me the importance of working hard. She works full-time and also makes sure our house is a clean, homey place. I do my best to help her around the house, and she always takes time to thank me and give me a peck on my cheek. I know my mother must be very tired sometimes, but she always find the strength to help me with my homework and ask me about my day. She meets with my teachers whenever I have a problem and, last year, she even volunteered to be an assistant coach on my soccer team. She is definitely my hero.

Part 4B: Student Writing Sample

48   D   W.PR.08.01 Writing Process
The title should reflect the main idea of the passage. Answer choice D, "A Fun Getaway," does this. The other answer choices give only details in the passage.

49   B   W.GN.08.01 Writing Genres
A short story is fiction; a myth is also fiction. This is not a biography about someone. It is closest to a memoir.

50   B   W.GR.08.01 Grammar and Usage
The author uses adjectives in lines 4 and 5. Answer choice B is the correct answer.

51   B   W.SP.08.01 Spelling
While "whether" is spelled correctly, it is not the correct spelling for the way the word is used.

52   B   W.SP.07.01 Spelling
The word "surrounded" has two r's, as in answer choice B.

Michigan Educational Assessment Program (MEAP)

W.PR. 08.01, .02, .03, .04, .05	Writing Process
W.GR.08.01	Grammar and Usage
W.SP.08.01	Spelling
W.HW.08.01	Handwriting

Writing: Peer Response to a Student Writing Sample
Grades 3–8

Holistic Scorepoint Descriptions

Here is an explanation of what readers think about as they score your writing.

4	The response clearly and fully addresses the task and demonstrates an understanding of the effective elements of writing that are relevant to the task. Ideas are supported by relevant, specific details from the student writing sample. There may be surface feature errors, but they do not interfere with meaning.
3	The response addresses the task and demonstrates some understanding of the effective elements of writing that are relevant to the task. Ideas are somewhat supported with a mix of general and specific relevant details from the student writing sample. There may be surface feature errors, but they do not interfere with meaning.
2	The response demonstrates limited ability to address the task and may show limited understanding of the effective elements of writing that are relevant to the task. Ideas may be supported with vague and/or partially relevant details from the student writing sample. There may be surface features that partially interfere with meaning.
1	The response demonstrates an attempt to address the task with little, if any, understanding of the effective elements of writing that are relevant to the task. The response may include generalizations about the student writing sample with few, if any, details. There may be surface feature errors that interfere with meaning.

53

Sample 4-point response: The author of "The Lake" probably lives in the city. He or she likes going to Uncle Ray's because it is by a lake that is very beautiful in the summer. The author likes to look at the tall trees and listen to the birds singing in the morning. The author's sister likes to watch the wildlife. This leads the reader to believe that Uncle Ray's house is very different from the author's home.

Reading

Lesson 1: Word Study

R.WS.08.01   Use word structure, sentence structure, and prediction to aid in decoding and understanding the meanings of words encountered in context.

R.WS.08.02   Use structural, syntactic, and semantic analysis to recognize unfamiliar words in context (idioms, analogies, metaphors, similes to infer, history of the English language, common word origins, syllabication).

R.WS.08.04   Know the meaning of frequently encountered words in written and oral contexts (research to support specific words).

R.WS.08.07   Use strategies (e.g., prior knowledge, text features, structures) and authentic content-related resources to determine the meaning of words and phrases in context (e.g., historical terms, content area vocabulary, literary terms).

Wondering About Words

"The two events happened **simultaneously**."
"No way!"
"It's the truth."
"You mean they happened at the same time?"
"Yes—Marcus was born right when the clock struck midnight on New Year's Eve last year!"
"That's so cool!"

These two speakers sound very excited about the events that happened *simultaneously*, but what exactly does *simultaneously* mean?

On the Michigan Educational Assessment Program (MEAP), you will be asked to choose the correct meaning of words in a reading passage. Some of these words will be familiar to you, and you will be able to easily choose the correct answer. For other words, however, you will have to figure out their meaning from the **context**. The context is how the word is used in the passage. When you are unsure of the meaning of a word, take a look at the words surrounding the word you don't know. They might hold valuable hints.

Also, note the author's **diction** (choice of words). What kind of words does the author use? What kind of picture do you see in your head when you read the word to yourself? Is it a positive or negative picture? This picture is called a **connotation**, and it can help you figure out the mystery word.

Breaking Words Down

Don't forget that knowing the meaning of prefixes can also help you. For example, take the word "bicycle." "Bi-" is a prefix that means "two," or "double." **Prefixes** are groups of letters that are added to the beginning of words that change, or alter, the meaning of the word.

Similarly, **suffixes** are just like prefixes, only they are added to the end of words. Take the longest word you just read in the conversation between the two excited speakers, "simultaneously." The suffix, "-ly," is added to the end of the **root word**, "simultaneous." A root word is a word that cannot be broken down any further without losing all meaning. The suffix "-ly" means "like" or "having similar qualities of."

But, what does "simultaneous" mean? If you remember what the bolded terms above mean, then you can put it together. Let's try this one together.

First, consider the context in which the word is used in the conversation. Note the words around the mystery word. The first speaker says, " . . . two events happened." So, we know more than one thing occurred. The second speaker asks the first speaker, "You mean they happened at the same time?" Then the other replies "Yes . . . " and lists the two events that happened at the same time.

By looking at the words surrounding the mystery word, you can find out the meaning of that word, just as we found out by looking at the conversation that the speakers were talking about two events that happened at the same time, or "simultaneously."

Some Words Mean Similar Things?

Believe it or not, some words that look very different from each other may have similar meanings. Words that aren't spelled the same but have similar meanings are called **synonyms**.

For example, the words "small" and "petite" look nothing alike. However, they are synonyms for each other, because both have similar meanings. Both refer to little things.

Similes

Finally, look for key words in the sentence that contains the unfamiliar word. Words such as "like" and "as" are **similes**. Similes help compare two different things by linking them together.

Read the statement below.

"That dog can walk like a circus elephant."

In voicing this statement, the speaker is comparing a certain dog's walking abilities to the way circus elephants walk. The simile "like" links the two very different subjects together by making a comparison.

Now read this statement.

"I'm as happy as a rabbit in a carrot farm."

We can see that "as" is also a simile, because it links two different subjects together by making a comparison. The human speaker is comparing himself to an animal near its favorite food.

Playing detective and asking yourself questions when reading can help you to understand unfamiliar words you may come across. You'll scratch your head in confusion less and nod your head more in understanding while having a more enjoyable time reading.

 # Activity

Try to figure out the meaning of these words from the way they are used in each sentence. Look the words up in a dictionary to check your answers.

- With seven siblings and twenty-six cousins, Matt had a *plethora* of relatives.

- All that remained of the small boat was some *flotsam* floating on top of the water.

- Shaking her head, Shelley admitted the location of her missing notebook was an unsolvable *conundrum*.

- Under the *tutelage* of my older sister Miranda, I finally learned to swim.

- Nan, Lil, Bob, and Ava are names that are examples of *palindromes*.

Passage 1

Read the following passage and answer the questions that follow. Use the tips in the question section to help you craft the best answer.

Bridging Michigan: The Mackinac Bridge

Michigan, consisting of two peninsulas, is one of the most uniquely shaped states in America. Prior to 1957, however, this shape posed a serious problem: How could travelers quickly and easily move from one peninsula to the other? The solution was the Mackinac Bridge, one of the longest and most well known bridges in the world. Known as "The Mighty Mac," the bridge spans five miles over Lake Michigan and Lake Huron to connect the two parts of the state.

Although the bridge was opened in 1957 and took only three years to build, the idea for the bridge goes back over a century. In the 1800s, people relied on boats to traverse the Straits of Mackinac, the narrow region at which Lake Michigan and Lake Huron meet. Using boats was not an easy solution, though, and providing that service year-round proved extremely difficult. In 1884, a newspaper declared that boats could not solve the problem and that a bridge or tunnel would be necessary.

At first, people scoffed at the thought. Building a bridge or digging a tunnel would require an immense financial investment, and there would be no guarantee that such a structure would be safe. Several events occurred around that time, however, that changed the critics' minds. In 1883, the famous Brooklyn Bridge in New York was opened to the public. This bridge totaled over a mile in length and would be expected to provide safe, efficient thoroughfare for many thousands of people each day.

In Michigan, some visionaries decided that a similar feat could be accomplished at the Straits of Mackinac. In 1884, a store owner copied a picture of the Brooklyn Bridge and challenged his patrons to imagine a similar structure in their town. A few years later, engineers in Scotland constructed the

Firth of Forth Bridge, which was far longer than even the Brooklyn Bridge. These structures proved that bridges, no matter how immense, were indeed feasible.

Another main motivation for the Mackinac Bridge came around 1888, when the Grand Hotel was built on nearby Mackinac Island. This renowned landmark brought an even greater call for a convenient way to cross the Straits.

Throughout the next several decades, inventors and engineers began envisioning creative solutions to the problem. One architect planned a complex zigzag of bridges and causeways that would cover over 17 miles of land in order to link the peninsulas. Government officials drew up plans for a multimillion-dollar ferry service. In 1920, a state official even proposed a floating tunnel! None of these imaginative proposals came to fruition, however. Most people agreed that building a traditional bridge would be the best strategy.

In the following years, supporters of a bridge made limited progress toward their goal. After the Great Depression, Michigan leaders formed the Mackinac Straits Bridge Authority. This group attempted to gain federal assistance in the building of the bridge, but the funding they had hoped for never materialized. The beginning of World War II in 1939 was another setback for the Authority. The federal government could not dedicate much money or attention to the project during the war.

However, during the war years, the Authority did invaluable preparation work. They performed extensive studies on the Straits of Mackinac, learning about its geology, weather patterns, and water currents. The researchers also observed the automobile and train traffic in the area to determine what demands would likely be placed on a bridge. In 1950, they consulted with the world's most experienced bridge engineers.

Despite another delay during the Korean conflict, the momentum of the bridge project was unstoppable. In 1953, investment companies and bankers began selling bonds to gather money for the project. The Michigan legislature also raised some tax money that would go toward paying for the maintenance of the bridge. Altogether, they raised over a hundred million dollars toward the project.

Once the funding had been divided and distributed, the next step was to begin building the superstructure. After a ceremony, the construction began on May 8, 1954. After generations of planning, the bridge was completed in just three years. Designed by Dr. David Steinman, the Mackinac Bridge totals five miles in length and is heralded as one of the world's longest suspension bridges.

 **Questions**

1   What does "came to fruition" mean in the sixth paragraph?

_____

_____

 **Tip**

Consider the meanings of the other words in that sentence. That will help you determine what "came to fruition" means.

2   Reread this paragraph from the passage:

Despite another delay during the Korean conflict, the momentum of the bridge project was unstoppable. In 1953, investment companies and bankers began selling bonds to gather money for the project. The Michigan legislature also raised some tax money that would go toward paying for the maintenance of the bridge.

What does the word "momentum" mean? How do you know?

_____

_____

 **Tip**

Read the paragraph carefully and think about the ideas it contains. What does it tell you about the bridge project?

3   The author writes, "Most people agreed that building a traditional bridge would be the best strategy." What does the word "strategy" MOST LIKELY mean?

_____

_____

 **Tip**

Think about what you've just read about the Mackinac Bridge. Many designs for the bridge were proposed, but in this sentence you learn that a traditional design was seen as the best strategy. What does "strategy" mean?

Now check your answers on the next page.

Passage 1: "Bridging Michigan: The Mackinac Bridge"

 **Answers**

1   **Sample answer:** The paragraph says that unusual designs never came to fruition, because people preferred the traditional bridge. Here, "came to fruition" would mean a design that was achieved.

2   **Sample answer:** In this paragraph, the word "momentum" means "power and influence." The project was delayed by conflict, but its momentum led it to successes.

3   **Sample answer:** A strategy is a plan of action, such as planning to build a bridge.

Passage 2

Read the following passage and answer the questions that follow. Use the tips in the question section to help you choose the best answer.

Excerpt from *The Old Curiosity Shop*
by Charles Dickens

Night is generally my time for walking. . . . One night I had roamed into the City, and was walking slowly on in my usual way, musing upon a great many things, when I . . . turned hastily round and found at my elbow a pretty little girl, who begged to be directed to a certain street at a considerable distance, and indeed in quite another quarter of the town.

"It is a very long way from here," said I, "my child."

"I know that, sir," she replied timidly. "I am afraid it is a very long way, for I came from there to-night."

"Alone?" said I, in some surprise.

"Oh, yes, I don't mind that, but I am a little frightened now, for I had lost my road."

"And what made you ask it of me? Suppose I should tell you wrong?"

"I am sure you will not do that," said the little creature, "you are such a very old gentleman, and walk so slow yourself."

I cannot describe how much I was impressed by this appeal and the energy with which it was made, which brought a tear into the child's clear eye, and made her slight figure tremble as she looked up into my face.

"Come," said I, "I'll take you there."

She put her hand in mind as confidingly as if she had known me from her cradle, and we trudged away together. . . . I observed that every now and then she stole a curious look at my face, as if to make quite sure that I was not deceiving her, and that these glances (very sharp and keen they were too) seemed to increase her confidence at every repetition. . . .

"Who has sent you so far by yourself?" said I.

"Someone who is very kind to me, sir."

"And what have you been doing?"

"That, I must not tell," said the child firmly.

There was something in the manner of this reply which caused me to look at the little creature with an involuntary expression of surprise; for I wondered what kind of errand it might be that occasioned her to be prepared for questioning. Her quick eye seemed to read my thoughts, for as it met mine she added that there was no harm in what she had been doing, but it was a great secret—a secret which she did not even know herself. . . . She walked on as before, growing more familiar with me as we proceeded and talking cheerfully by the way, but she said no more about her home. . . . It was not until we arrived in the street itself that she knew where we were. Clapping her hands with pleasure and running on before me for a short distance, my little acquaintance stopped at a door and remained on the step till I came up knocked at it when I joined her. . . . There was a noise as if some person were moving inside, and at length a faint light appeared through the glass [and] an old man with long grey hair, whose face and figure as he held the light above his head and looked before him as he approached, I could plainly see. . . . The place through which he made his way at leisure was one of those receptacles for old and curious things which seem to crouch in odd corners of this town and to hide their musty treasures from the public eye in jealousy and distrust. There were suits of mail standing like ghosts in armour here and there, fantastic carvings . . . rusty weapons of various kinds, distorted figures in china and wood and iron and ivory: tapestry and strange furniture that might have been designed in dreams. . . .

[T]he child addressed him as grandfather, and told him the little story of our companionship.

"Why, bless thee, child," said the old man, patting her on the head, "how couldst thou miss thy way? What if I had lost thee, Nell!"

"I would have found my way back to YOU, grandfather," said the child boldly; "never fear."

The old man kissed her, then turning to me and begging me to walk in, I did so. The door was closed and locked. Preceding me with the light, he led me through the place I had already seen from without, into a small sitting-room behind, in which was another door opening into a kind of closet, where I saw a little bed that a fairy might have slept in, it looked so very small and was so prettily arranged. The child took a candle and tripped into this little room, leaving the old man and me together.

"You must be tired, sir," said he as he placed a chair near the fire, "how can I thank you?"

"By taking more care of your grandchild another time, my good friend," I replied.

"More care!" said the old man in a shrill voice, "more care of Nelly! Why, who ever loved a child as I love Nell?"

He said this with such evident surprise that I was perplexed what answer to make. . . . I was surprised to see the child standing patiently by with a cloak upon her arm, and in her hand a hat, and stick.

"Those are not mine, my dear," said I.

"No," returned the child, "they are grandfather's."

"But he is not going out to-night."

. . . "Oh, yes, he is," said the child, with a smile.

"And what becomes of you, my pretty one?"

"Me! I stay here of course. I always do."

I looked in astonishment towards the old man. . . . Alone! In that gloomy place all the long, dreary night.

[She] cheerfully helped the old man with his cloak, and when he was ready took a candle to light us out. . . . [The old man] walked on at a slow pace. . . . I remained standing on the spot where he had left me, unwilling to depart. . . . [I] could not tear myself away, thinking of all possible harm that might happen to the child. . . . I continued to pace the street for two long hours; at length the rain began to descend heavily, and then over-powered by fatigue though no less interested than I had been at first, I engaged the nearest coach and so got home. A cheerful fire was blazing on the hearth, the lamp burnt brightly, my clock received me with its old familiar welcome; everything was quiet, warm and cheering, and in happy contrast to the gloom and darkness I had quitted.

But all that night, waking or in my sleep, the same thoughts recurred and the same images retained possession of my brain. I had ever before me the old dark murky rooms—the gaunt suits of mail with their ghostly silent air—the faces all awry, grinning from wood and stone—the dust and rust and worm that lives in wood—and alone in the midst of all this lumber and decay and ugly age, the beautiful child in her gentle slumber, smiling through her light and sunny dreams.

 **Questions**

1   Read this sentence from the passage:

There was something in the manner of this reply which caused me to look at the little creature with an involuntary expression of surprise; for I wondered what kind of errand it might be that occasioned her to be prepared for questioning.

In this sentence, the word "involuntary" means

A   frightening.
B   unintentional.
C   disgusted.
D   bored.

 **Tip**

Picture the look of surprise on the narrator's face as he wondered what the little girl had been doing on the streets at night all alone. How might surprise affect a person's facial expression? How does he feel about this little girl? If you aren't sure, go back and read this part of the story.

2   At the end of the story, the narrator pictures the girl "alone in the midst of all this lumber and decay and ugly age." The word "midst" means

A   house.
B   dust.
C   middle.
D   things.

Tip

Try to see in your mind what the narrator is seeing in his. Where is the little girl? Substitute each of the answer choices for the word "midst" to see which one makes the most sense.

3   Shortly after meeting the little girl, the narrator tells the reader that he "wondered what kind of errand it might be that occasioned her to be prepared for questioning." As it is used here, the word "occasioned" means

A   parted.
B   caused.
C   suffered.
D   wanted.

Tip

Find this sentence in the passage and read this section of the story. What is the narrator wondering about the little girl? Which answer choice makes the most sense?

Now check your answers on the next page.

Passage 2: "An Excerpt from *The Old Curiosity Shop*"

 **Answers**

1   B   The story does not suggest that the narrator's look frightens the little girl, nor does it say anything that would lead the reader to believe he was disgusted. He is certainly not bored by the little girl's secrecy. You can tell that a *voluntary* action is intentional, or chosen, intentional because to volunteer means to do something by choice; therefore, an action that is *involuntary* is an unintentional action. The narrator does not mean to look surprised, but people who are surprised often reveal emotions that they did not mean to show.

2   C   The narrator pictures the little girl sleeping in the middle of all of the strange objects and "musty treasures" he saw when the old man let him into the shop. Answer choice C makes the most sense.

3   B   The narrator wonders what the little girl has been doing, and what caused her to reply that her activities were a secret. This is the only answer choice that makes sense in the context of the story.

Passage 3

Read this passage and answer the questions that follow. Use the tips under each question to help you choose the best answer.

Amazing Morocco

Morocco is a land of sights and smells. In fact, it's hard to describe Morocco without painting a picture filled with vibrant reds, oranges, and yellows. Add the heady aromas of spices, and you'll begin to get a sense of this vibrant country. Everywhere you turn, your senses are nearly overwhelmed with sights and sounds and smells.

The colorful country of Morocco, once known as Mauritania, is located in northern Africa, touching both the Atlantic Ocean and the Mediterranean Sea. It also shares borders with the countries of Algeria and Western Sahara. If you look on a map, you might be surprised to see that a section of Morocco's shoreline nearly touches Spain. Morocco's closeness to Spain is reflected in much of its culture. France has had a great deal of influence on Morocco as well. The country is also the third most populous Arab country in the world, behind Egypt and Sudan, and this fact plays a major role in Morocco's culture and its people's everyday lives.

Most Moroccans speak Arabic, and many educated people in the cities speak French and sometimes Spanish. The original Moroccans were Berbers, a non-Arab African tribe of people who have subsisted in the area for thousands of years. Nearly ninety percent of Moroccans today are either entirely or partially of Berber descent. Arabs are the second largest group of people living within Morocco and have influenced much of Moroccan culture, including religion. There are also nearly 100,000 French people living in Morocco.

For many years, the Berbers roamed in the Moroccan countryside, speaking their own language and passing their history from generation to generation by telling stories. This oral tradition is still kept alive by small bands of musicians that travel the country and perform at weddings and other celebrations. Today, many Berbers are mixed into the Arab mainstream culture. Though some Berbers still live in rural areas raising livestock and growing crops, many are educated and have jobs in the cities.

If you travel to Morocco from Spain, you will take just a short boat ride to the city of Tangier. This closeness to Spain is what gives Tangier its nickname, the Gateway to Morocco. The tales behind the founding of Tangier fascinate outsiders. According to some legends, Tangier was founded by the fabled Hercules

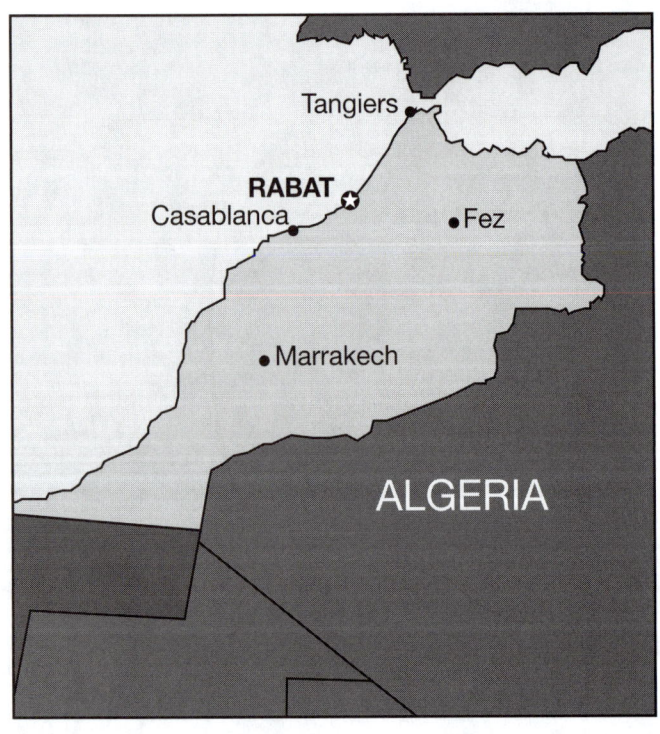

of Greek mythology, who slept in a cave just outside the city while attempting one of his 12 labors. This story has made the cave one of Tangier's most popular tourist attractions. This coastal city is also known for its breathtaking views, and is the first stop in Morocco for many tourists from all over the world.

As you travel inland, you'll arrive at the historic city of Fez. This city was built on the crossroads of old caravan paths. Fez is a very religious city and carries on the great traditions of Islam. The city of Fez has been Morocco's capital in the past, the last time carrying the title until 1912 when the country came under French rule. It is the home of treasured mosques, which are the churches of the Islamic religion.

Traveling down the Atlantic coast, you'll come to the city of Rabat, which is the capital of Morocco. Rabat was the capital of Morocco under French rule and remains the capital of the country. Rabat is the official residence of the Moroccan king. Rabat is also home to a large university and a school of music, dance, and theater arts. Rabat is connected to its sister city Salé by a bridge over the river that separates the cities.

Just a little farther down the Atlantic coast of Morocco is the city of Casablanca, which is the largest city in Morocco. Casablanca is the business capital of Morocco and is a large center for foreign trade and banking. While the French were in control of Casablanca, they built the largest man-made harbor, which helped the Moroccans to buy and sell products with Europe. Casablanca is also home to the world's second largest mosque, built in honor of the former Moroccan King Hassan II. The city gained international fame from the 1942 film *Casablanca* featuring actors Humphrey Bogart and Ingrid Bergman as star-crossed lovers during the early days of World War II.

The southern part of Morocco is home to the city of Marrakech, located in the foothills of the Atlas Mountains. Nowhere are the colors and smells of Morocco stronger and more powerful than in Marrakech. Many merchants sell handmade goods here and the leather products and woven carpets are among the finest in the world. Arabs and Berbers alike come to sell their goods and sip mint tea in the blazing sunlight.

With its colorful culture, Morocco has been the inspiration for films like *Casablanca* and *Lawrence of Arabia*, famous artists like French painter Eugene Delacroix, and books like Edith Wharton's *In Morocco*. While the country's exotic blend of old and new has inspired people from across the globe, Morocco's own artists have made a significant impression on the world as well. Painter Farid Belkahia was well known for using natural pigments to paint ancestral symbols in his art, and Ahmed Cherkaoui used Arabic calligraphy and Berber symbols in many of his works.

Life in Morocco is a mix of old traditions and modern ways. Within Moroccan cities, there are the remaining old cities, called medinas. Medina streets are narrow mazes that connect the gated courtyards of private homes with the central marketplaces. Many courtyards are filled with trees and flowers and fountains. Moroccan people living in these old sections of the cities use their courtyards for sewing, weaving, eating meals, and visiting neighbors. You'll find many children in the narrow streets carrying water to their homes or bread to be baked in the ovens down the street. The marketplaces of Morocco are filled with strange and amazing sights. Besides the merchants selling their carpets and spices, there are dancers and musicians, acrobats and magicians, and even snake charmers.

 Questions

1   The author writes, ". . . it's hard to describe Morocco without painting a picture filled with vibrant reds, oranges, and yellows." The word "vibrant" means

A   pale.
B   clear.
C   bright.
D   mixed.

 **Tip**

What point is the author trying to make about Morocco? Reread the first paragraph if you're not sure.

2   The following sentence appears in the third paragraph of the article:

The original Moroccans were Berbers, a non-Arab African tribe of people who have subsisted in the area for thousands of years.

Based on how the word is used in the article, which of the following BEST describes the meaning of "subsisted"?

A   lived
B   grown
C   traveled
D   breathed

 **Tip**

Eliminate answer choices that don't fit with the author's point. Which is the best match?

3   Read the following excerpt from the passage:

Add the heady aromas of spices, and you'll begin to get a sense of this vibrant country. Everywhere you turn, your senses are nearly overwhelmed with sights and sounds and smells.

In this excerpt, the word "aromas" means

A   tastes.
B   smells.
C   sounds.
D   pictures.

 **Tip**

Consider the words that surround "aromas." Do they give you any clues as to the meaning of this word?

Now check your answers on the next page.

Passage 3: "Amazing Morocco"

 **Answers**

1   C   If the reader is unsure of the meaning of the word, he or she may be able to figure out the meaning by looking at the words that surround it. By looking at the sentence in the passage, the reader can easily see that the meaning of the word "vibrant" is bright.

2   A   By looking at the sentence, the reader can see that the word "subsisted" means "lived." Choice A is correct.

3   B   "Aromas" means "smells." Spices don't make sounds, and having pictures of spices doesn't seem right. The colors were discussed earlier. Spices are known for having strong smells. So, answer choice B must be correct.

Passage 4

Read this passage and answer the questions that follow. Use the tips under each question to help you choose the best answer.

The Common Cold: Unbeatable Bug?

Cold season is upon us again, and it seems like we're all searching for a miracle cure. While the human race has accomplished many amazing feats, curing the common cold has not been one of them. Even after hundreds of years of research, scientists cannot stomp out the tiny bug.

Nearly everyone knows the miserable feeling of having a cold. The sickness can affect your entire head with symptoms including sneezing, coughing, a blocked or runny nose, a sore throat, and a headache. Some especially nasty colds carry along all of these symptoms.

A cold can really disrupt your life. Every year, millions of people miss work or school due to colds. This hurts the economy and the educational system. People spend billions of dollars each year for medicines to reduce their nagging symptoms. So far, no medicine has been proven beyond a doubt to work, however.

The term "cold" is actually just a label for a group of symptoms. These symptoms are actually caused by viruses, microscopic parasites that rely on other organisms to live. When these viruses find a good host—you, for instance—they thrive, but end up giving you some sort of disease in return. Talk about ungrateful guests!

However, in the case of a cold, the disease itself isn't usually what bothers you. The most annoying symptoms are actually caused by your own immune system, which does everything it can to eradicate the virus. The immune system will produce extra mucus to try to flush out the virus through the nose or to trap it and take it to the stomach, where it will be killed with stomach acid. This mucus, of course, results in the cold's most infamous symptom: the runny nose. The stuffy head symptom is caused by inflammation of blood vessels above the nose. Sneezing and coughing are brought about by the irritation caused by the immune system's defense mechanisms. Coughing, in particular, occurs when the irritation moves to the lungs.

Perhaps the only good thing about this cycle of discomfort is that it usually results in success within a week or so. The virus dies and is flushed away, and good health returns. Most people, however, still lament the fact that colds last as long as they do.

Throughout history, people have tried hundreds of different remedies in order to shorten the durations of their cold symptoms. Even powerful leaders like Napoleon Bonaparte concerned themselves with their nasal health. Napoleon had a recipe for cold pills that he claimed worked wonders for him—this recipe included such peculiar ingredients as "ipecacuanha root," "squill root," "gum ammoniac," and "gum arabic." Other old-time recipes called for almonds, currants, poppy seeds, or licorice.

Remedies like these may sound funny today, but we have to be humble because we haven't found a surefire remedy, either. Many scientists today believe that certain kinds of zinc, a substance often used in vitamins, provide our best hope for relief. Chicken soup is always a safe bet, too.

Scientists agree that the best way to avoid the nuisance of a cold is to avoid catching the cold in the first place. Cold viruses are transmitted from person to person. Contrary to popular belief, colds are not caused by cold weather. Going outside on a wintry day will not make you any more likely to catch a cold. There are several different reasons that people usually get colds during the colder months. The main reason is that, in the winter, more people stay indoors and end up in close contact with one another.

The virus is usually transmitted by way of people's hands. That transfer doesn't only happen when people shake hands. It can also happen through objects that are touched by several people. For instance, if someone coughs on his or her hand and then touches a doorknob, the next person to touch the doorknob may pick up a virus. This transfer is practically unavoidable, so the best thing to do is simply to wash your hands. Also, you can try not to touch your nose or eyes, because they're easy landing zones for cold viruses, which take up residence in the cells inside the nose.

Although there are ways to reduce your exposure to colds, some scientists believe Americans suffer through as many as a billion colds per year. There's no relief in sight. For now, the common cold remains an unbeatable bug.

 Questions

1   The following sentence appears in the fifth paragraph of the article:

The most annoying symptoms are actually caused by your own immune system, which does everything it can to eradicate the virus.

Based on how the word is used in this article, which of the following BEST describes the meaning of "eradicate"?

A   to get rid of
B   to surround
C   to create with
D   to absorb

 Tip

If you don't know what this word means, read the sentence carefully. Think about what the article said about the immune system's role in dealing with viruses. What does your immune system want to do with the virus?

2   The author writes that "the cold's most infamous symptom" is the runny nose. The word "infamous" means

A   painful.
B   well-known.
C   long-lasting.
D   bothersome.

 Tip

Go back to the passage and read the sentences around this one. What point is the author trying to make?

3   Read this sentence:

Most people, however, still lament the fact that colds last as long as they do.

In this sentence, the word "lament" means

A   express.
B   complain.
C   will not accept.
D   will not believe.

Tip

Think about how it feels to have a cold. What do people sometimes do when they have a cold for a long time?

4   The author writes that people still "haven't found a surefire remedy" for the common cold. The word "surefire" means

A   medication.
B   vaccination.
C   readily available.
D   completely effective.

Tip

Read over the sentence carefully. If you're still unsure of the meaning of the word, look back to the passage for hints.

Now check your answers on the next page.

Passage 4: "The Common Cold: Unbeatable Bug?"

 Answers

1 A The word "eradicate," as used in the sample sentence, means "to get rid of something." The immune system does not want to surround, create, or absorb the virus. It wants to eradicate it—get rid of it—completely.

2 B The word "infamous" means "well-known." The runny nose is the symptom most commonly associated with the common cold. While the answer choice "bothersome" also seems to be correct, this is not the point the author is trying to make.

3 B "Lament," in this instance, means "complain." People complain about how long the cold lasts. While answer choice D might also be acceptable, answer choice B is the better answer.

4 D A "surefire remedy" is completely effective and would get rid of all of the cold's symptoms. Answer choice D is the best answer. While some of the other answer choices may look good, they don't fit the context as well as answer choice D.

 # Lesson 2: Narrative Text

R.NT.08.01   Investigate through classic and contemporary literature recognized for quality and literary merit various examples of distortion and stereotypes such as those associated with gender, race, culture, age, class, religion, and other individual differences.

R.NT.08.02   Analyze elements and style of narrative genres (e.g., historical fiction, science fiction, realistic fiction).

R.NT.08.03   Analyze the role of rising and falling actions, minor characters in relation to conflict, and credibility of the narrator.

R.NT.08.04   Analyze how authors use symbolism, imagery, and consistency to develop credible narrators, rising and falling actions, and minor characters.

Questions About Narrative Texts

Narrative texts are passages that tell a story. There are many elements and techniques involved in narrative texts. Questions about narrative texts might ask you to identify the conflict or the tone of a short story. They might ask you to analyze how the themes in the story relate to your own life.

Questions dealing with literary elements and techniques may ask you to analyze the characters in a story. Who are they, and what are their behaviors, personalities, and motivations? Who is the protagonist, or central character of the story? Who is the antagonist, the character or force that opposes the central character?

You might be asked to identify or analyze writing conventions (how the author uses writing to convey meaning). For example, a question might ask you why an author repeats a certain idea or why an author uses a specific technique, such as symbolism. You will also likely be asked questions about genres (like historical fiction, realistic fiction, and science fiction).

Activity

In a sentence or two, summarize the plot of a story that matches each of the literary genres below. It can be a story you've read or one you made up. The first has been done for you.

Literary Genre:	Example Plot:
1. Myth	A cursed king turns everything he touches into gold.
2. Comedy	
3. Tragedy	
4. Historical fiction	
5. Science fiction	
6. Biography	

Passage 1

Read this passage and answer the questions that follow.

New Problems for New Jake

More than anything, Jake envied the popular kids in his class. They had all the newest clothes, the best haircuts, and the most friends. He decided that the best thing he could do to improve himself would be to become one of the popular kids.

Jake decided to reinvent his image. He gathered his lawn-mowing money and went into town. Over the next few hours he bankrupted himself, but when he returned home, he had fancy clothes and a trendy haircut. He was eager to go to school the next day.

At school, Jake surprised many people with his new look. Some of the students seemed to notice him for the first time. However, their reactions weren't as positive as he had hoped. Most of the kids seemed to view him with curiosity or confusion. Nobody raced up to Jake, complimenting him or asking to be his friend.

The popular kids did not flock to him either. A few muttered to one another, "Is that the same Jake who used to sit in the corner?" Then they ignored him.

Jake was puzzled that things didn't seem to have changed much. At lunch, be bought a tray of food and decided to sit at the table where all the most popular kids sat. When the students at the table saw him approaching, they muttered to one another. Some snickered, and Jake began to sweat. He went to an empty part of the table and sat down, trying to smile at the table's occupants. They looked at him but did not smile back.

Now Jake was too distressed to eat, and he just stared at his tray. He watched the steam from his spaghetti fade. When lunch period was finally over, he was relieved to get out of the cafeteria.

While the students were walking back to class, Jake spotted Darren, the star of the football team and one of the big shots in the school social scene. Darren was one of the people who had snickered at him.

"What do you and your friends have against me?" asked Jake. "I'm willing to dress like you, talk like you, and act like you, and you don't even give me the smallest bit of respect in return."

Darren looked at him with surprise. "That's the point," he said. "You're not you anymore. We all liked you just fine until you started pretending you were someone else."

Jake realized then that his life was never perfect, but it was pretty good until he started dressing up and putting on acts. The next day he looked and acted like he used to. He was better off just being himself.

 **Questions**

1   What is Jake's plan for "reinventing" himself?

   _____

   _____

2   What does Jake learn at the end of the story? Does this lesson change his behavior?

   _____

   _____

3   How would you describe Darren? Do you think he'll be a friend to Jake?

   _____

   _____

Now check your answers on the next page.

Passage 1: "New Problems for New Jake"

 **Answers**

1   **Sample answer:** Jake hopes to change and improve his life by becoming popular at school. He thinks he can accomplish this by purchasing fancy clothing and getting a trendy haircut.

2   **Sample answer:** By the end of the story, Jake learns that people respected him for being himself but did not respect him when he tried to act like others. He stops dressing like the popular kids and returns to acting like his old self.

3   **Sample answer:** Darren is one of the most popular students at the school, but he is not a likable character at first. He snickers at and shuns Jake. However, after Jake confronts him, Darren speaks honestly to Jake. The two will probably be on friendly terms after their talk.

Passages 2 and 3

Read the following poems. Then answer the questions that follow.

Because I could not stop for Death
by Emily Dickinson

Because I could not stop for Death—
He kindly stopped for me—
The Carriage held but just Ourselves—
And Immortality.

We slowly drove—He knew no haste
And I had put away
My labor and my leisure too,
For His Civility—

We passed the School, where Children strove
At Recess—in the Ring—
We passed the Fields of Gazing Grain—
We passed the Setting Sun—

Or rather—He passed us—
The Dews drew quivering and chill—
For only Gossamer, my Gown—
My Tippet—only Tulle—

We paused before a House that seemed
A Swelling of the Ground—
The Roof was scarcely visible—
The Cornice—in the Ground—

Since then—'tis Centuries—and yet
Feels shorter than the Day
I first surmised the Horses' Heads
Were toward Eternity—

I heard a Fly buzz—when I died
by Emily Dickinson

I heard a Fly buzz—when I died—
The Stillness in the Room
Was like the Stillness in the Air—
Between the Heaves of Storm—

The Eyes around—had wrung them dry—
And Breaths were gathering firm
For the last Onset—when the King
Be witnessed—in the Room—

I willed my Keepsakes—Signed away
What portion of me be
Assignable—and then it was
There interposed a Fly—

With Blue—uncertain stumbling Buzz—
Between the light—and me—
And then the Windows failed—and then
I could not see to see—

 Questions

Questions 1 through 3 refer to the poem "Because I could not stop for Death."

1　The author begins the poem, "Because I could not stop for Death." Based on this line, you can tell that

　A　the woman is afraid to die.
　B　the woman wants to die.
　C　the woman did not expect to die.
　D　the woman is already dead.

 Tip

This question asks you to understand the feelings of the protagonist, based on the way she words the first line of the poem. Reread the first stanza of the poem. Why do you think she could not stop for Death?

2   A word that can be used to describe Death in this poem is

   A   lovable.
   B   scary.
   C   kind.
   D   mean.

Tip

Go back to the poem and find the first time the narrator mentions Death. Study the words she uses to describe Death and Death's actions.

3   The woman tells her story

   A   a long time after the events happen.
   B   before the events happen.
   C   shortly after the events happen.
   D   as the events happen.

Tip

Study the tense of the verbs used in the poem. Also, reread the last stanza of the poem for clues about the woman's place in time.

Questions 4 and 5 refer to the poem "I heard a Fly buzz—when I died."

4   When the speaker says that the stillness in the room was like the stillness in the air "Between the Heaves of Storm," she means that

   A   it was about to become noisy in the room.
   B   it was probably going to rain outside.
   C   something bad is about to happen.
   D   the people in the room are angry.

Tip

Think about how people feel when they know a storm is approaching. How is this like what the people in the poem are about to witness?

5   What does the King symbolize?

    A   rain

    B   death

    C   a fly

    D   a friend

 **Tip**

Think about what is happening to the speaker. What happens to her at the end of the poem?

Questions 6 and 7 refer to both "Because I could not stop for Death" and "I heard a Fly buzz—when I died."

6   In BOTH poems, the speakers

    A   are surrounded by family.

    B   are discussing their death.

    C   view death in a positive way.

    D   hear a strange noise.

 **Tip**

Reread both poems. What are they about?

7   In what way is the speaker of "Because I could not stop for Death" DIFFERENT from the speaker of "I heard a Fly buzz—when I died"?

    A   She sees many things before death.

    B   She knows her death is approaching.

    C   She gives away her things.

    D   She looks out a window.

 **Tip**

Begin by eliminating answer choices that apply to the speaker of "I heard a Fly buzz—when I died."

Now check your answers on the next page.

Passages 2 and 3: "Because I could not stop for Death" and "I heard a Fly buzz—when I died"

 **Answers**

1   C   While several of the answer choices seem as if they are correct, only one answer choice shows the expressed feelings of the protagonist: answer choice C. The woman, the speaker of the poem, says that she was too busy to stop for Death but that Death kindly stopped for her.

2   C   The answer to this question is stated in the poem. The narrator says that Death "kindly stopped" for her.

3   A   You need to reread the end of the poem to find this answer. She states, "Since then—'tis Centuries," meaning centuries have passed since these events took place.

4   C   The people standing around the speaker are anticipating her death. The speaker means that something bad is about to happen.

5   B   After the speaker says that people were waiting for the last onset when the King be witnessed, she dies. Therefore, the King is a symbol for death.

6   B   The speakers in both poems discuss their death. While answer choice C might also seem correct, we don't know for sure that the speaker in the second poem views death positively. Answer choice B is the best answer.

7   A   The speaker in the first poem goes on a carriage ride and sees many things after Death comes for her. The speaker in the second poem sees only what is nearby.

Passage 4

Many legends exist about the Babylonian hero King Gilgamesh. Read this adaptation of one of his adventures. Then answer the questions that follow.

The Myth of King Gilgamesh

Thousands of years ago, the people of ancient Babylonia had great respect for their king, Gilgamesh. He was the son of powerful gods and had superhuman strength. Gilgamesh had keen intelligence and great foresight. He oversaw the construction of Uruk, a beautiful city.

However, Gilgamesh had many flaws. He was arrogant and brash, and frequently neglected the needs of his people in favor of his own desires. He was also oppressive and demanded complete control of everyone in his kingdom. When he began to interfere in people's weddings, the people of Babylonia decided that something had to be done. They flocked to their temple and prayed to their chief god, Anu, pleading with him to confront Gilgamesh and end his exploitation. Their prayers were answered with silence, however, and they left the temple disappointedly.

The next day, a hunter named Shuja headed into the forests outside of the city in search of game. As soon as he stepped into the thick, shadowy woods, he heard the roar of an animal he did not recognize. It resembled a horrifying combination of the growls, hoots, whistles, and barks of a dozen different species. He heard it again, and it was closer this time. Before he could flee, he found himself face-to-face with a hulking wild man surrounded by a team of vicious animals.

An hour later, an exhausted Shuja returned to the city. He looked so ragged and terrified that a crowd gathered around him, inquiring what troubles had befallen him. "I encountered a wild man in the forest training animals for warfare," Shuja explained. "His name was Enkidu, and he said Anu had dispatched him to dethrone King Gilgamesh."

A worried murmur passed through the crowd. *What would happen if such a menacing creature attacked Uruk?* they wondered. The prospect was even less pleasant than living under Gilgamesh's continued oppression. They realized they needed to stop Enkidu, but how could they negotiate with an animal-like man? Some thought they should fight. Others thought they should flee. Still others thought they should surrender to the creature and try to reason with it. Nobody could agree on a course of action.

"Stop this quarrelling. I'll get us out of this predicament," announced Shamhat, one of the most beautiful women in Uruk. The next morning she left the city's protective walls and proceeded into the forest in search of Enkidu. She found him at a watering hole where he and his supporters had stopped to rest. Shamhat approached him confidently, and he could sense that she was not motivated by apprehension or hostility. This caught Enkidu off guard.

Shamhat addressed Enkidu with kindness and compassion, and he responded in a similarly civil manner. They spent the day together and, the next morning, she led him into Uruk as a friend, not an enemy. The people gathered around them and celebrated the cessation of his threat. Enkidu, though disoriented by the new environment, came to love the beauty, companionship, and sophistication he encountered inside the city walls. Taking up residence with some shepherds, Enkidu learned how to behave like a civilized human being.

Meanwhile, Gilgamesh had been having visions of powerful, mysterious newcomers trespassing upon his land. It was therefore no surprise to him to learn of Enkidu's presence in Uruk. Gilgamesh consulted with his mother, who advised him to embrace this newcomer as a friend, because together they were destined for great accomplishments.

What does she know? Gilgamesh thought bitterly. *I would not degrade myself by accepting some wild man as a companion.*

And so Gilgamesh continued his oppression of the people. During a marriage celebration, Gilgamesh interfered again. He was jealous of the groom and intended to kidnap the bride. He believed he was justified in doing so because he was the ruler of Uruk, and he was comfortable with the knowledge that nobody would challenge him. But he had forgotten about the newcomer, Enkidu, who suddenly appeared in the king's doorway and refused to allow him to break up the wedding.

"How dare you exploit your people for your own gain!" demanded Enkidu.

"How dare you question my decisions!" roared Gilgamesh, lunging forward to attack his challenger.

The two combatants struggled for hours, their powers equally balanced. Finally, Gilgamesh was able to secure an advantage by raising a sword high over Enkidu. Instead of bringing the sword slashing down, however, he paused and then slowly lowered the weapon.

"You are a worthy opponent," he admitted, "and I was wrong to belittle you. I see the wisdom in your challenge, and I will not spoil the wedding." Gilgamesh helped Enkidu to his feet, and they shook hands. "I think my mother was right. If you and I work together, we can accomplish great things for the people of Uruk."

Questions

1   Gilgamesh began to respect Enkidu after Gilgamesh

   A   decided to ignore his mother.
   B   explored Enkidu's past.
   C   learned that Enkidu was a mighty warrior.
   D   asked the people for their advice.

Tip

   Skim over the text, looking for the moment that Gilgamesh began to respect Enkidu. Study what happened right before then.

2   The people of Babylon were upset with King Gilgamesh because he

   A   forbade them to pray to Anu.
   B   listened to his mother's advice.
   C   made friends with a wild man.
   D   interfered in wedding ceremonies.

Tip

   Reread the text, looking for clues on why the people get upset with King Gilgamesh.

3   The author records Gilgamesh's thoughts after he speaks with his mother in order to

   A   show that Gilgamesh is keeping secrets.
   B   contrast Gilgamesh's attitude with his mother's.
   C   suggest that Gilgamesh is afraid to speak.
   D   explain Gilgamesh's immediate reaction.

Tip

   Reread Gilgamesh's thoughts after he speaks with his mother. Think about why the
author would include these in the story.

4   Based on the way the author tells the story, this text would be BEST described as a

A   myth.
B   memoir.
C   drama.
D   poem.

 **Tip**

Think about what details usually make a story fit into a certain genre. Then think about this story's details.

5   What is one theme of this story?

A   Treat others as you want to be treated.
B   Always listen to your mother.
C   Do not show weakness in battle.
D   Do not let greed destroy you.

 **Tip**

Reread the story, paying special attention to Shamhat's actions.

6   Gilgamesh didn't want to make friends with Enkidu because

A   he was afraid of Enkidu.
B   he knew Enkidu wanted to be king.
C   his mother warned against it.
D   he thought he was better than Enkidu.

 **Tip**

Reread the section of the story when Gilgamesh decides not to be friends with Enkidu. Think about what makes him come to this decision.

7   How did Enkidu change after meeting Shamhat?

    A   from peace-loving to warlike

    B   from brave to cowardly

    C   from untamed to mannerly

    D   from noble to selfish

 **Tip**

Compare Enkidu's interaction with Shuja with the way Enkidu interacts with Shamhat.

Now check your answers on the next page.

Passage 4: "The Myth of King Gilgamesh"

 Answers

1   C   Gilgamesh began to respect Enkidu after he fought him. Answer choice C is correct.

2   D   King Gilgamesh had his ups and downs, but the biggest problem he caused the people of Babylon was that he interfered in wedding ceremonies. The story said he felt justified in kidnapping brides because he was the king and nobody could stop him.

3   B   The author records Gilgamesh's thoughts to show that he does not agree with his mother.

4   A   This question asks you the genre of this story. This story can best be described as myth because of the exaggerated strength and abilities of the characters. This story also lacks some of the conventions of the other genres in the list.

5   A   Shamhat treats Enkidu with kindness and he is friendly in return.

6   D   Gilgamesh thought he was much better than Enkidu. This is why he did not want to be friends with him. Answer choice D is correct.

7   C   When Shuja first meets Enkidu, Enkidu is wild. When Shamhat treats Enkidu kindly, he begins to act more mannerly. Answer choice C correct.

Passage 5

Read this excerpt about a woman who is determined to find out something. Then answer the questions that follow.

Mrs. Rachel Lynde Is Surprised
An Excerpt from *Anne of Green Gables*
by Lucy Maud Montgomery

There are plenty of people in Avonlea and out of it, who can attend closely to their neighbor's business by dint of neglecting their own; but Mrs. Rachel Lynde was one of those capable creatures who can manage their own concerns and those of other folks into the bargain. She was a notable housewife; her work was always done and well done; she "ran" the Sewing Circle, helped run the Sunday-school, and was the strongest prop of the Church Aid Society and Foreign Missions Auxiliary. Yet with all this Mrs. Rachel found abundant time to sit for hours at her kitchen window, knitting "cotton warp" quilts—she had knitted sixteen of them, as Avonlea housekeepers were wont to tell in awed voices—and keeping a sharp eye on the main road that crossed the hollow and wound up the steep red hill beyond . . .

She was sitting there one afternoon in early June. The sun was coming in at the window warm and bright; the orchard on the slope below the house was in a bridal flush of pinky-white bloom, hummed over by a myriad of bees. Thomas Lynde—a meek little man whom Avonlea people called "Rachel Lynde's husband"—was sowing his late turnip seed on the hill field beyond the barn; and Matthew Cuthbert ought to have been sowing his on the big red brook field away over by Green Gables. Mrs. Rachel knew that he ought because she had heard him tell Peter Morrison the evening before in William J. Blair's store over at Carmody that he meant to sow his turnip seed the next afternoon. Peter had asked him, of course, for Matthew Cuthbert had never been known to volunteer information about anything in his whole life.

And yet here was Matthew Cuthbert, at half-past three on the afternoon of a busy day, placidly driving over the hollow and up the hill; moreover, he wore a white collar and his best suit of clothes, which was plain proof that he was going out of Avonlea; and he had the buggy and the sorrel mare, which betokened that he was going a considerable distance. Now, where was Matthew Cuthbert going and why was he going there?

Had it been any other man in Avonlea, Mrs. Rachel, deftly putting this and that together, might have given a pretty good guess as to both questions. But Matthew so rarely went from home that it must be something pressing and unusual which was taking him; he was the shyest man alive and hated to have to go among strangers or to any place where he might have to talk. Matthew, dressed up with a white collar and driving in a buggy, was something that didn't happen often. Mrs. Rachel, ponder as she might, could make nothing of it and her afternoon's enjoyment was spoiled.

"I'll just step over to Green Gables after tea and find out from Marilla where he's gone and why," the worthy woman finally concluded. . . .

Accordingly after tea Mrs. Rachel set out; she had not far to go; the big, rambling, orchard-embowered house where the Cuthberts lived was a scant quarter of a mile up the road from Lynde's Hollow. To be sure, the long lane made it a good deal further. Matthew Cuthbert's father, as shy and silent as his son after him, had got as far away as he possibly could from his fellow men without actually retreating into the woods when he founded his homestead.

Green Gables was built at the furthest edge of his cleared land and there it was to this day, barely visible from the main road along which all the other Avonlea houses were so sociably situated. Mrs. Rachel Lynde did not call living in such a place LIVING at all. . . .

Mrs. Rachel rapped smartly at the kitchen door and stepped in when bidden to do so. The kitchen at Green Gables was a cheerful apartment—or would have been cheerful if it had not been so painfully clean as to give it something of the appearance of an unused parlor. Its windows looked east and west; through the west one, looking out on the back yard, came a flood of mellow June sunlight; but the east one, whence you got a glimpse of the bloom white cherry-trees in the left orchard and nodding, slender birches down in the hollow by the brook, was greened over by a tangle of vines. Here sat Marilla Cuthbert, when she sat at all, always slightly distrustful of sunshine, which seemed to her too dancing and irresponsible a thing for a world which was meant to be taken seriously; and here she sat now, knitting, and the table behind her was laid for supper.

Mrs. Rachel, before she had fairly closed the door, had taken a mental note of everything that was on that table. There were three plates laid, so that Marilla must be expecting some one home with Matthew to tea; but the dishes were everyday dishes and there was only crab-apple preserves and one kind of cake, so that the expected company could not be any particular company. Yet what of Matthew's white collar and the sorrel mare? Mrs. Rachel was getting fairly dizzy with this unusual mystery about quiet, unmysterious Green Gables.

 **Questions**

1   What word BEST describes Mrs. Lynde?

   A   caring
   B   nosy
   C   busy
   D   worried

 **Tip**

Reread the text. Pay special attention to Mrs. Lynde's comments and actions. What do they say about her?

2   Mrs. Lynde PROBABLY takes "a mental note of everything" on the table at the Cuthberts' because

   A   she wants to know if she'll be invited for tea.
   B   she plans to gossip about their messy house.
   C   she is trying to figure out who is coming to their house.
   D   she wants to put down their choice of dinnerware.

 **Tip**

Reread the end of the story. Think about why Mrs. Lynde is in the house.

3   The last sentence of the fourth paragraph says that Mrs. Lynde's "afternoon enjoyment was spoiled." Her afternoon was spoiled because

   A   she was worried about her neighbor.
   B   she did not know everything that was going on.
   C   she knew she had to visit Marilla Cuthbert.
   D   she could not spend time at her window.

 **Tip**

Reread this paragraph. Think about what you know about Mrs. Lynde.

4   The Cuthberts PROBABLY live so far from the main road because

   A   people in Avonlea did not like them.
   B   Matthew Cuthbert's father made them.
   C   they did not like the sound of traffic.
   D   they liked their privacy.

 **Tip**

Reread the story, thinking about which of the above reasons is most likely.

5   Why is it odd that Matthew Cuthbert is leaving Green Gables?

   A   because his field needs to be sown
   B   because he doesn't like to leave town
   C   because he is wearing a white collar
   D   because Marilla is making dinner

 **Tip**

Reread this section of the text, paying attention to details.

Now check your answers on the next page.

Passage 5: "Mrs. Rachel Lynde Is Surprised"

 **Answers**

1 B The beginning of the passage does say that Mrs. Lynde was busy, but she is mostly nosy. She looks out her window watching others and becomes upset because she does not know what Matthew Cuthbert is doing. Answer choice B is the correct answer.

2 C Mrs. Rachel is trying to figure out who is coming to their house because she is trying to make sense of what she believes to be Matthew Cuthbert's odd behavior.

3 B The main reason that Mrs. Lynde's afternoon is spoiled is that she does not know where Matthew has gone. This is the best answer.

4 D As it says in the passage, Matthew Cuthbert's father decided to locate the house there because he was as shy as his son. With such a nosy neighbor, it makes sense to live so far away. Answer choice D is the best answer.

5 B Matthew Cuthbert's behavior is unusual because Matthew seldom leaves the house. He is quiet and shy and does not like socializing with strangers. Answer choice B is correct.

Lesson 3: Informational Text and Critical Standards

R.IT.08.01   Analyze elements and style of informational genre (e.g., comparative essays, newspaper writing, technical writing, persuasive essay).

R.IT.08.02   Analyze organizational patterns (e.g., theory, evidence, sequence).

R.IT.08.03   Explain how authors use text features to enhance the understanding of central, key, and supporting ideas (e.g., illustrations, author pages, prefaces, marginal notes).

R.CS.08.01   Evaluate the appropriateness of shared, individual, and expert standards based on purpose, context, and audience in order to assess their own work and work of others.

Questions About Informational Text and Critical Standards

Authors write for many reasons. They might write an article that gives readers information or teaches them how to do something. They might write a brochure to make people aware of a product. Authors sometimes write articles or letters to convince readers to feel as they do or to persuade readers to take a certain action.

Questions about informational text might ask you to identify the purpose of a piece of writing. You might have to decide if a passage is meant to inform, instruct, entertain, or persuade. You might be asked what an author intended by including examples or diagrams in the passage. Usually, these devices are used to make something complex easier to understand. You might also be asked why an author has chosen to include a certain detail in a passage. To do this, you will have to identify the author's purpose for writing the passage and how including these details might help the author achieve that purpose.

You may have to use a passage's organization to find answers. Authors organize their writing for certain reasons. For instance, some authors write to compare and contrast ideas or characters. Sometimes these similarities or differences are stated in the passage. Other times you have to read the passage carefully to determine the ways in which two ideas or characters are alike or different.

Read this letter and then follow the instructions on the next page.

Dear Lindsay,

I had so much fun on my trip!

First of all, the drive to Muskegon was a nightmare! John picked on me the whole time, pulling my hair and calling me names. On the way back, Mom said he had to be nice to me so he let me pick the CDs we listened to all the way home. But I'm getting ahead of myself. I wanted to tell you about what I did at the Michigan Adventure and Wild Water Park.

When we first got there, we had to park a million miles away and catch a tram to the entrance of the park. Then we had to wait in line before we could buy tickets. From the line, I could hear the roller coasters zooming down the tracks as well as the screams of the passengers. I was so excited. I couldn't wait to go inside.

Once we got in, John wanted to ride the Hydroblaster, but Mom didn't want us to get wet yet, so we didn't ride that until right before we went to the water park. Instead we started off on the Corkscrew, and what a beginning that was! John pretended like he was going to get sick when we were getting off, but he was just fine when we got over to Shivering Timbers. Wooden roller coasters are the best! I love the noise they make.

John said his favorite ride was the Lazy River, which was boring if you ask me. It was in the water park, where we didn't go to until a few hours after lunch. In the water park, I had the most fun on the Snake Pit. That was the very first ride we went on when we got to the water park, and, in my opinion, the best.

My favorite is still the Wolverine Wildcat though. Nothing has ever matched the time when you and I rode the Wolverine three times in a row. That is a memory that I'll never forget.

Can't wait to see you!

Josie

Each box contains one event from Josie's letter. Write the event in the box where it belongs on the timeline.

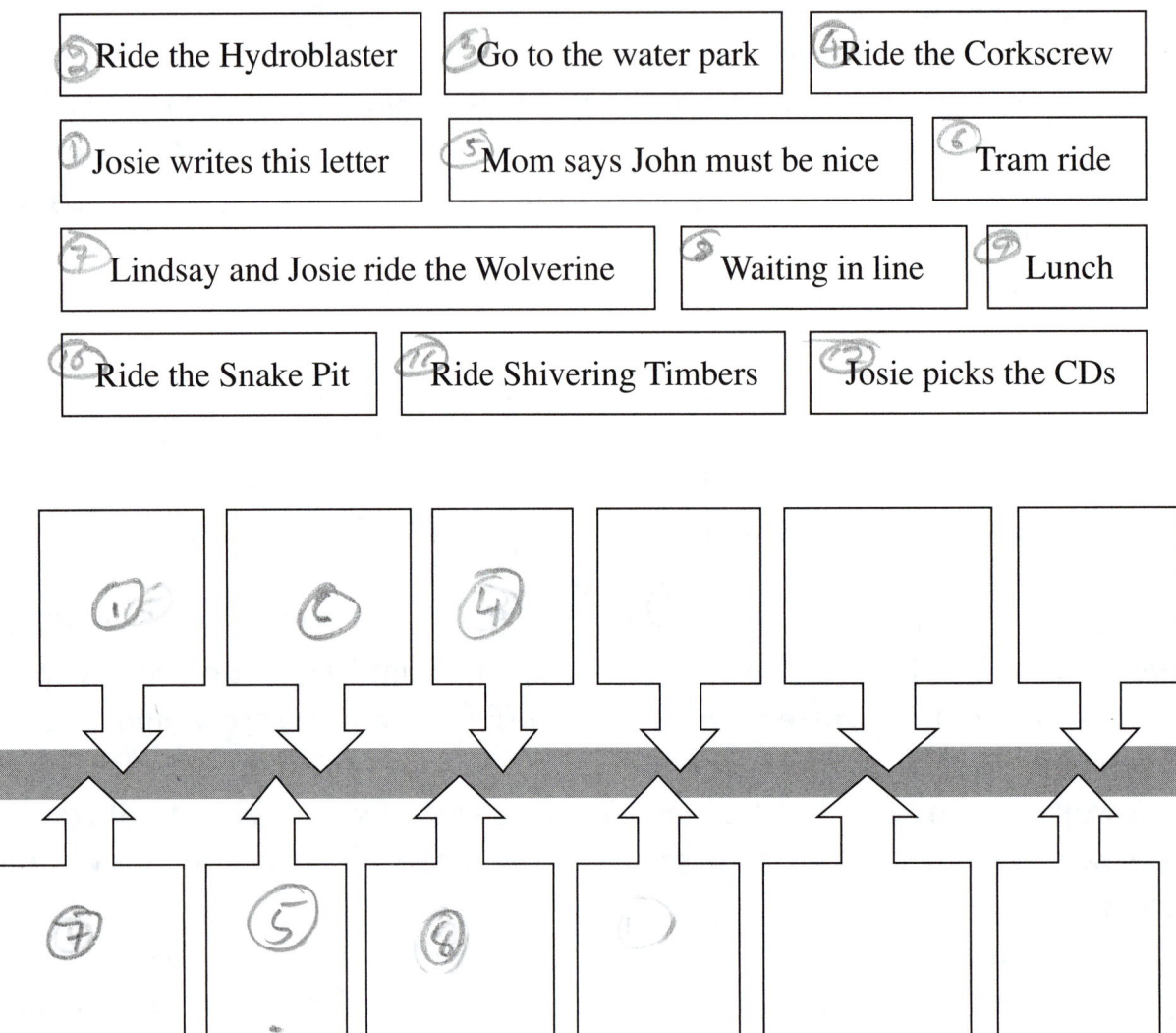

Ride the Hydroblaster Go to the water park Ride the Corkscrew

Josie writes this letter Mom says John must be nice Tram ride

Lindsay and Josie ride the Wolverine Waiting in line Lunch

Ride the Snake Pit Ride Shivering Timbers Josie picks the CDs

Passage 1

Read this passage. When you finish reading the passage, answer the questions that follow.

Say Goodbye to the Cleavers

The Way We Were?

Households today are very different from those of the past. In the 1950s, most American families tried to model themselves after the seemingly perfect Cleavers on the television show *Leave It to Beaver*. The vast majority of homes had a stay-at-home mother, who cleaned, cooked, and cared for the children, and a father who was in charge of discipline and the family finances. Fathers were said to have a "newspaper for a face" since they preferred the business section to interaction with their offspring, unless their authority was needed. Most families had many children, and mothers strove to keep their homes as spotless and "Cleaver-like" as humanly possible. Many families sat down together for a home-cooked meal each night and often had a weekly menu plan that might serve meatloaf on a particular night and fried chicken on another.

The problem with the "Cleaver ideal" is that it leaves little room for diversity, meaning not every family fits happily into this mold. Though some households in the 1950s may have resembled, or tried to resemble, the Cleavers' pleasant existence, the truth is many families did not have this type of lifestyle, and families today are looking less and less like the Cleavers.

Women in the Workforce

Though June Cleaver may have enjoyed baking cookies and doing laundry all day, many women today are not content to spend their days taking care of the household. In the 1950s, women were expected to get married, have children, and take care of the house. However, when more and more women began attending college in the 1960s and '70s, jobs that were once considered "men's work," like positions in the fields of business and medicine, started opening up to women. This represented a shift in society's idea of what women could and were expected to do. Though it might have been strange for a mother to have a job in the '50s, today it is very common. Today, more than sixty percent of families in the United States have a mother who is employed outside of the home. The real difference between the women of the 1950s and the women of today is that now women can choose to work or to stay home, and women of the "June Cleaver" era usually didn't have any options.

Mr. Mom

Many modern fathers are much more involved in child-rearing than their predecessors. Their roles extend far beyond discipline and include hugs, kisses, and diaper changes. In fact, in 2003 the U.S. Census reported that about 105,000 fathers are "stay-at-home" dads who have taken over the role that was traditionally held by mothers to raise their young children while their partners go to work. Even if they don't stay home with their children, dads today often help out around the house, cook for the family, and are involved in their children's school and extracurricular activities.

Breaking the Mold

Of course, not all families are "nuclear" like the Cleavers, meaning that they include a mom, a dad, and their children. Some kids live with their grandparents, an aunt or uncle, or just one of their parents.

Single-parent households have always existed, but today they are becoming more and more common. Some children have parents who are divorced, and they spend a certain amount of time with one parent and then go stay with the other parent for a while. Other kids live with just their mom or their dad. While a two-parent, dual-income household may be the most common type of household today, it is certainly not the only one. According to the census, there are 2 million single fathers and about 10 million single mothers in the United States. It's easy to see that fitting the modern family into an old-fashioned stereotype is like trying to put a square peg into a round hole. Americans do not want to be the Cleavers anymore.

 Questions

1   One way in which families today differ from those of the past is that

    A   families today have more children.
    B   families today share a meal each night.
    C   most mothers today are in charge of disciplining.
    D   most mothers today work outside of the home.

 **Tip**

Reread the passage if you're not sure of the answer. Eliminate answer choices that are clearly incorrect.

2   What does the author say is wrong with the "Cleaver ideal" of the past?

    A   Families today do not fit into this mold.
    B   The Cleavers were a TV family.
    C   The Cleavers did not include a mother and a father.
    D   Families today do not want to let go of the past.

 **Tip**

This question asks you to identify what is wrong with the "Cleaver ideal." Go back to the passage and find the place where the author discusses this. Think about all the different types of families you have seen in your own life. What makes this ideal wrong for modern families?

3   According to the passage "Fitting the modern family into an old-fashioned stereotype is like trying to put a square peg into a round hole." When the author makes this comparison, he or she is trying to say that families today

    A   are more square than round.

    B   are more rounded than before.

    C   care more about making money.

    D   don't fit the same mold they did before.

 **Tip**

This question asks you to explain a comparison the author makes in the passage. Be careful answering this question. First, read through all the answer choices. Then, eliminate the ones that are not relevant to the question.

Now check your answers on the next page.

Passage 1: "Say Goodbye to the Cleavers"

 **Answers**

1   D   The answer to this question is stated in the passage. The passage does not say that families today have more children, share a meal each night, or have fathers who are in charge of discipline. These are traits of families of the past. Answer choice D is correct: most families today have a mother who works outside of the home.

2   A   In the passage, the author says that the "Cleaver ideal" leaves little room for diversity and that not many families today fit into this mold. While it is true that the Cleavers were a TV family, this does not explain why the ideal does not work for families today (choice B). Choice C is not true, and though the "Cleaver ideal" still exists (choice D), it doesn't explain why it doesn't work. The logical choice is answer A.

3   D   This question asks you what the author means when he compares modern families with an old-fashioned stereotype. Think about the stereotypes and how they often don't apply to modern families. Answer choice D is the best answer choice.

Passages 2 and 3

Read the passages "The Day of the Dead" and "Make Your Own Bread of the Dead" and answer the questions that follow.

The Day of the Dead

People in Mexico celebrate a special holiday, called the Day of the Dead, in which they honor loved ones who have passed away and celebrate life for the living. This holiday actually began more than three thousand years ago, when a large group of people lived in Mesoamerica, which includes what is now known as Mexico. In this agricultural society, life and death were considered seasonal, since crops sprouted, grew, bore fruit, and then died. The people had a similar seasonal view of human life and death and celebrated it with a month-long festival. The Mesoamericans felt that the dead returned to visit their living relatives and friends during the unique celebration.

The Mesoamericans often used skulls as decorations during this celebration. While this might seem frightening and even gruesome, they considered it a natural approach to celebrate life. Masks called *caretas* were worn by dancers to scare away the spirits of the dead at the conclusion of the event.

Today, people living in this area gather near the end of October to begin preparations for the Day of the Dead celebrations. Houses and businesses are decorated with banners, which are hung across the streets. Banners and other decorations are usually in black and white and accented with pink, yellow, and gold. Skeletal images made out of papier-mâché are assembled to look as if they watch people from the windows of storefronts. Families build *ofrendas*, or altars, on which to place beautiful items to attract the souls of the departed. They place photographs and items that the deceased relatives or friends cherished during their lives. Families also place flowers, fruits, candles, and incense on the altars.

Traditionally, special marigolds called *zempasúchil* are used as decorations during this celebration along with incense called *copal*. Both the flowers and the incense are strong smelling and people believe that these strong smells help the spirits find their way back after the festivities.

Relatives prepare extravagant dishes that were favorites of their loved ones. A traditional food often prepared at this time is *mole*, which is a thick sauce made from many ingredients, including many spices, chilies, sesame seeds, and various herbs, as well as chocolate and fruit. *Mole* takes a long time to make, so it's considered a special gift for cherished loved ones.

Once families believe that the spirits of their loved ones have finished eating, the food offerings are eaten or given away. Food and beverages are a big part of the Day of the Dead celebrations. Families believe that the spirits are tired and thirsty after the long journey. A special drink called *atole*, which is made from corn meal, water, and flavorings, is prepared and offered to quench the thirst of the dead. A glass of water is offered as well.

Some celebrate with sugar candies in the shape of skulls. Another type of candy that's shaped into death figures is marzipan. Along with the sugary marzipan, there's also a traditional pumpkin candy made from large green pumpkins that are grown just for the Day of the Dead celebration. Another specialty of this time of year is the *pan de muerto*, or bread of the dead. Breads are often round with extra dough placed on top in the shape of bones. Breads are also made in the shapes of humans and animals, especially rabbits. Some bakeries have to hire extra helpers to make enough bread for this special time of the year.

Family members go to the cemeteries where they clean and mend the gravesites. Graves are decorated with lavish flower arrangements and candles. On the day of the celebration, it's almost a party-like atmosphere, with people wandering about visiting the elaborately decorated graves. In some places, at six o'clock in the evening, the bells begin to ring to summon the dead. *Mariachis*, or local musicians, play the favorite melodies of the deceased. People spread out blankets and enjoy foods that were favorites of their loved ones. People stay the whole night with multitudes of candles burning, and priests make their rounds and pray with the families.

The Day of the Dead has a vibrant historical background that resonates through thousands of years. Though it's celebrated today in diverse ways, the Day of the Dead brings families together. Today, some families celebrate with a simple family dinner featuring the bread of the dead. In many traditional families, each family member plays a role in more elaborate preparations. The Day of the Dead is a time when families honor their relatives and friends by retelling their favorite stories, eating their favorite foods, and singing their favorite songs, which keeps their cherished loved ones alive in their hearts.

Make Your Own Bread of the Dead

The Mexican Day of the Dead may sound like a sad event, but in actuality it's a time for feast and celebration. One of the most famous aspects of the Day of the Dead is the delicious food. Sugar candies, chili, special drinks, and even a unique kind of bread grace tables throughout Mexico on this special day. The bread of choice is called *pan de muerto*—the bread of the dead. Despite its somewhat eerie name, *pan de muerto* is a delicious, eagerly anticipated dish.

The recipe used for *pan de muerto* varies from area to area, and it has also changed through history. Many bakers today use modern ingredients, such as condensed milk, to make the bread richer and sweeter. However, the basic ingredients remain relatively standard. *Pan de muerto* is a food you can easily make in your own kitchen, whether your kitchen is in Mexico, America, or anywhere else.

To make bread of the dead, you will need the following ingredients:

Pan de muerto

- $\frac{1}{2}$ cup milk

- $\frac{1}{2}$ cup water

- $\frac{1}{2}$ cup butter

- 5 $\frac{1}{2}$ cups flour

- 2 packages dry yeast

- 1 teaspoon salt

- 1 tablespoon whole anise seed

- $\frac{1}{2}$ cup sugar

- 4 eggs

To make a special glaze for the bread, you will need these additional ingredients:

Glaze

- $\frac{1}{3}$ cup orange juice

- $\frac{1}{2}$ cup sugar

- 2 tablespoons orange zest

To make the bread:

1. Heat the milk, the water, and the butter in a small saucepan.
2. Combine 1 $\frac{1}{2}$ cups of the flour with the yeast, the salt, the anise seed, and the sugar.
3. Mix the warmed contents of the saucepan with the flour mixture. Beat until the ingredients are combined.
4. Add the eggs and beat them into the mixture.
5. Continue adding the rest of the flour until the dough is soft.
6. Knead the dough until smooth and stretchy. This may take five to ten minutes.
7. Place the dough in a greased bowl and cover it with plastic stretch wrap. Leave it in a warm spot for about 1 $\frac{1}{2}$ hours. After that, it should have risen to about twice its original size.
8. Preheat your oven to 350 degrees Fahrenheit.
9. Knead the dough again and shape it into round loaves. You can sculpt these loaves into skull shapes or animal shapes, or leave them round and arrange strips of dough (shaped like bones) on top of them.
10. Bake the dough for 35 to 40 minutes.

To make and apply the glaze:

1. Mix the juice, the sugar, and the zest in a saucepan. Set it on heat and bring to a boil for two minutes.

2. Use a pastry brush to apply the glaze to the bread.

 Questions

Questions 1 and 2 refer to "The Day of the Dead."

1   The information in "The Day of the Dead" would be MOST useful for a student research paper on

A   Mexican-American ancestors.

B   early methods of agriculture.

C   musical preferences in Mexico.

D   traditional beliefs about life and death.

 Tip

Look over the answer choices and eliminate those that are clearly not right. Then, think about the information in the first selection. What did you learn by reading it?

2   Why do people prepare *atole* during the celebration?

A   Their relatives liked to drink *atole* when they were alive.

B   The dead are thirsty and drink *atole* to quench their thirst.

C   The strong smell will help the dead find their way back home.

D   It takes a long time to make and is therefore considered special.

 Tip

Scan the selection for information about *atole*. Then, skim that information. What is the main purpose of *atole*?

Question 3 refers to "Make Your Own Bread of the Dead."

3   People who read "Make Your Own Bread of the Dead" will learn how to

    A   prepare a special drink.

    B   apply glaze to food.

    C   string up banners.

    D   make papier-mâché.

 **Tip**

    Think about what the selection is mostly about. What information is the author trying to provide?

Questions 4 and 5 refer to both "The Day of the Dead" and "Make Your Own Bread of the Dead."

4   Based on the two selections, which statement is true?

    A   *Pan de muerto* is not meant to be eaten by the living.

    B   Many special foods are prepared for the Day of the Dead.

    C   The Day of the Dead is a solemn time for mourning the dead.

    D   Only special bakeries in Mexico can prepare *pan de muerto*.

 **Tip**

    Look back to the selections, and then read the answer choices carefully. Which statement is supported by both selections?

5   Which of the following do BOTH passages include?

    A   the many ingredients of *pan de muerto*

    B   the best method to keep dough from sticking

    C   a description of the symbolic meaning of skulls

    D   a description of Day of the Dead celebrations

 **Tip**

    Look over these answer choices carefully. All of these choices refer to information in the passages, but only one item is included in both passages. Which is it?

Now check your answers on the next page.

Passages 2 and 3: "The Day of the Dead" and "Make Your Own Bread of the Dead"

 Answers

1   D   The article mentions most of these topics, but it does not go into nearly enough detail to assist a student writing a paper. The only topic that it does cover in detail is the traditional view of life and death in Mexico. Answer choice D is best.

2   B   In the passage, it says that many people believed that spirits had to make long journeys during the festivals. Because of this, the people prepared *atole* drinks to quench the spirits' thirst.

3   B   After reading the passage, you should know how to apply the sugar and orange glaze to the *pan de muerto*. The other topics are not mentioned in this passage.

4   B   Three of these answer choices are incorrect. The bread is meant to be eaten by the living, and it, along with other foods, can be baked in just about any kitchen. The Day of the Dead is not solemn, but is celebratory. The correct choice is that many special foods are prepared for the events.

5   D   Answer choices A and B are mentioned in the second selection, and answer choice C is mentioned in the first selection. However, both selections discuss answer choice D. That is the correct answer.

Passage 4

Read this passage about the world's largest horse. Then answer the questions that follow.

A Horse from History

What do Renaissance Italy and modern-day Grand Rapids, Michigan, have in common? This may sound like the beginning of a very strange riddle. However, it only becomes stranger when the answer is revealed: The connection between these two different times and places is *Il Cavallo*, the world's largest horse!

Il Cavallo began long ago in the mind of the remarkable Renaissance artist Leonardo da Vinci and now resides in Frederik Meijer Botanical and Sculpture Gardens in Grand Rapids. The Sculpture Gardens are open to the public—you can stop by and say hello to the horse!

The History

Over five hundred years ago, an Italian duke, Ludovico Sforza, hired Leonardo da Vinci to tackle an incredible project. Sforza wanted Leonardo to construct the world's largest statue of a horse. This enormous equine would be cast in bronze and would stand, twenty-four feet tall, near his palatial home in the city of Milan.

Leonardo took up one of his notebooks and sketched a tiny drawing of a horse. This first drawing was only the size of a postage stamp, but Leonardo already realized that he loved the idea. The horse would be a symbol of strength, beauty, and peace. He set about studying and sketching horses, planning one of the most striking statues of all time.

After countless hours of preparation, Leonardo began constructing a full-size model of the horse out of clay. This was the first step in the process of casting a bronze statue; however, the rest of the process would never take place. During a war, French soldiers captured the palace of Duke Sforza. When they saw the huge clay horse, they drew their crossbows and began firing arrows at it. They used the clay model for target practice until it was damaged beyond repair.

Leonardo's work was ruined, and his dream was lost. Some stories hold that he regretted this for the rest of his life, loathing the mention of it because its memory was too painful. Some lore suggests he even wept on his deathbed over the loss of the horse. Regardless, Leonardo's *Il Cavallo* was never completed, and the entire idea went nearly forgotten for centuries.

Rebirth of the Dream

In 1978, a pilot and sculptor named Charles C. Dent read an article about Leonardo's vision for *Il Cavallo*—a name which simply means "The Horse." Dent was fascinated and quickly began an endeavor to sculpt a replica of the horse using Leonardo's blueprints. Then, as his enthusiasm grew, Dent began making plans to have a full-size statue constructed. Not only did he want to make Leonardo's vision a reality, but he also wanted to honor the people and history of Italy. He would make the horse a gift to them.

This would be a giant project, and Dent knew he couldn't do it alone. He helped to gather a foundation of people willing to donate time and money to the lofty goal. When Dent died in 1994, the foundation continued his work and invested millions of dollars and thousands of hours into the project. Scholars, artists, metal specialists, sponsors, and even animal experts worked together to develop a statue that would do Leonardo honor.

Building a Reality

After the workers had made their plans, they built an eight-foot model of the horse they hoped to build. Then they used special enlarging machines to increase the size of the horse to a huge twenty-four feet. Finally, the most demanding part of the project was handed over to workers at a foundry, or metal-working shop. They put in many months of patient labor to cast the approximately sixty small bronze parts of the giant statue.

The bronze was cast in thin sheets that would serve as the statue's skin. Since they were too thin to hold up the weight of the assembled statue, the foundry workers had to reinforce the sheets with a heavy "skeleton" underneath them. Later still, the workers would have to attach the sheets to the supports, and then join them all together in a beautiful and seamless fashion.

When they completed this, however, the result was astonishing: a fifteen-ton bronze steed, taller than a house! As soon as it was done, thousands of people flocked to see it. This horse was not meant to be kept in its stable, however; it was instead shipped to Milan, Italy. In 1999, *Il Cavallo* was set up in a cultural park for all the people of the city to enjoy. It stood right where Leonardo and Sforza might have placed theirs, if it hadn't been destroyed by war.

But the story doesn't end there. Seeing that so much work had been invested in the single statue, some people decided to make a second one using the same molds and tools. This identical twin horse would be kept in America—Frederik Meijer Gardens in Grand Rapids, Michigan, to be exact. On October 7, 1999, this *Il Cavallo* was unveiled as part of a joyous ceremony in which bands, jugglers, and Renaissance re-enactors celebrated the new life brought to the dreams of Leonardo da Vinci. Leonardo's horse, the impressive symbol of peace and strength, would ride again!

Quick *Il Cavallo* Facts

Height: 24 feet

Weight: 15 tons

Composed of silicon bronze, with skeleton of stainless steel

Construction began around 1980

Unveiled on October 7, 1999

Location: Frederik Meijer Botanical and Sculpture Gardens, Grand Rapids, Michigan

Stands near an educational center that includes an identical *Il Cavallo* in 1:3 scale

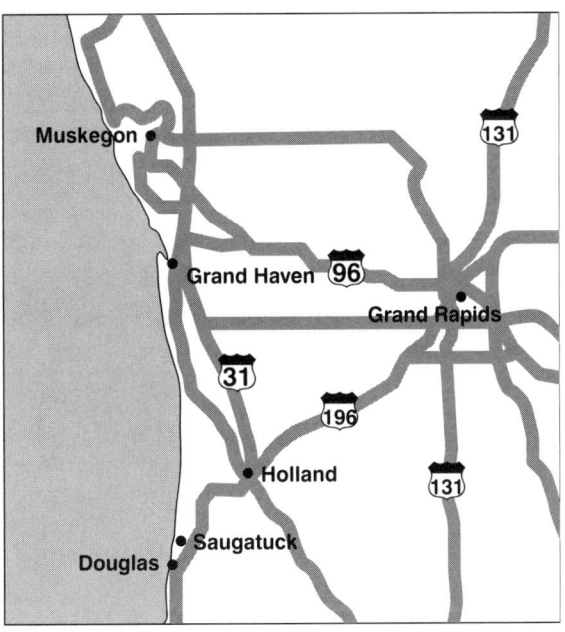

Grand Rapids is easily accessible from many major highways

 **Questions**

1 This passage is an example of

   A   a letter.
   B   a brochure.
   C   a research report.
   D   a dictionary entry.

 **Tip**

Think about the above kinds of writing. Which one best fits the style of "A Horse from History"?

2 One of the two *Il Cavallo* statues was given to Italy because

   A   Italy wanted a special gift from America.
   B   most of the people who built the statues were Italian.
   C   it would be displayed in Leonardo's hometown.
   D   Americans wanted to honor Italy's history and people.

 **Tip**

If you don't remember what caused one statue to be given away, look back to the text.

3 This passage is MOST LIKELY intended for an audience made up of

   A   experts.
   B   tourists.
   C   kindergarten students.
   D   horse lovers.

 **Tip**

Think about the information in the passage, and the way it is presented. Is the material easy or difficult? Who would probably want to read it?

4   The author PROBABLY wrote this passage in order to

   A   encourage people to visit *Il Cavallo*.
   B   ask people to donate money to the project.
   C   entertain readers with a story about a big bronze horse.
   D   help Americans and Italians understand one another.

Tip

Think about what the author is trying to accomplish in this passage. What words does he or she choose? What tone does he or she take?

5   What is the MOST STRIKING difference between Leonardo's first drawing and the finished statue?

   A   The drawing was lost for centuries, while the statue will never be lost.
   B   The drawing was as small as a stamp, while the statue is 24 feet tall.
   C   The drawing was made carefully, while the statue was built carelessly.
   D   The drawing was very colorful, while the statue is only one color.

Tip

If you don't remember what made the drawing and the final statue very different, look back to the passage.

6   The map was MOST LIKELY included in this passage in order to

   A   help people find Grand Rapids, Michigan.
   B   teach visitors about highways.
   C   let people calculate how many miles they need to drive.
   D   show visitors where *Il Cavallo* is located in the Meijer Gardens.

Tip

Look carefully at the map. What information does it show you?

Now check your answers on the next page.

Passage 3: "A Horse from History"

 **Answers**

1   B   This passage contains information, including a short history, a map, and a fact box, that would likely be of most help to a tourist. The passage is most similar to a brochure.

2   D   The passage explains that Charles Dent wanted to give a gift to Italy to celebrate its people and history. The correct answer is D.

3   B   The information here would probably be most helpful to a tourist, who might need a map to find Grand Rapids and some background information on the statue there.

4   A   The author seems excited about *Il Cavallo* and wants people to see it. Phrases like "you can stop by and say hello to the horse!" invite people to visit.

5   B   There can be many differences between a drawing and a statue. However, the passage says that Leonardo's first drawing was only the size of a stamp. That is very small, especially when compared with the huge statue that resulted. Answer choice B is best.

6   A   The map does not show a scale of miles, or the exact location of Meijer Gardens. It does show Michigan's highways, but that's not the most important part. Clues from the text and the caption inform you that Grand Rapids is the focus of this map.

 # Lesson 4: Comprehension

R.CM.08.01   Connect personal knowledge, experience, and understanding of the world to themes and perspectives in the text.

R.CM.08.02   Read, retell, and summarize grade-level appropriate narrative and informational texts.

R.CM.08.03   State global themes, universal truths, and principles within and across texts to create a deeper understanding.

R.CM.08.04   Apply significant knowledge from what has been read in grade-level appropriate science and social studies texts.

What Is Comprehension?

From the moment you look at words on a page, you begin to understand their meaning. The way words look forms an impression on you. For example, if you look at a reading passage with dense paragraphs and detailed diagrams, you might understand that the passage is difficult and complex. You would probably approach the reading passage differently than you would if you were reading a comic strip.

How well you understand what you read depends on your experience. For example, if you have read many complex reading passages, you might know that breaking the passage into smaller parts will make it easier to understand. Other factors also contribute to how well you understand what you read. Your mood can affect your understanding. And if you're preoccupied—if something else is on your mind—it will also affect your understanding. The trick is to train yourself to stay focused and aware of what you're reading.

Some questions on MEAP are designed to test your reading comprehension and metacognition skills. The word **comprehension** simply means "understanding." When you comprehend a piece of writing, you understand what you have read.

On MEAP, questions on comprehension might ask you to summarize a passage or explain its most important points. You might be asked the author's purpose for writing the passage. You might be asked to use the ideas you bring to the text—your prior experience, knowledge, and beliefs—to explain your particular understanding of a passage. For fictional passages, you might be asked to identify a major theme of the passage.

Activity

Spend a few moments looking at this photograph. Then cover the photo and answer the questions on the next page without looking back at the photo.

Grand Circus Park

 **Questions**

1   What is the title of the photo?

_____ Grand Circus Park _____

2   In the photo, are there cars on the street?

_____ Yes there are cars _____

3   Is the picture of the past or the present?

_____ Past _____

4   What does the sign say on the Dime Savings Bank?

_____ 100,000 people, Dime savings bank _____

How Did You Do?

Because you knew you would be answering questions about the photo, you probably looked at the picture differently. Maybe you paid more attention to detail. Congratulations! Formulating an efficient way to study is using your metacognitive skills.

Passage 1

Think about the why the author wrote this passage as you read. Then answer the questions that follow.

Healthy Earth Hotels

People travel to escape the stresses of their daily lives, and when they visit a hotel, they expect to be able to relax in a clean, comfortable environment. Many hotels go to great lengths to pamper their guests; sometimes, however, these efforts can become excessive. For instance, hotels regularly wash bathroom towels and bed sheets every day. This leads to unnecessary usage of time, money, and resources, as well as damage to the environment. For this reason, the Healthy Earth Hotel program should be adopted by inns throughout America.

The Healthy Earth Hotel program is elementary, but highly effective. In order to implement this program, hotel operators can request a simple start-up kit that includes sturdy laminated signs and specially designed laundry hangers. These signs and hangers, when installed in each guest room, quickly and conveniently educate guests about the Healthy Earth Hotel plan and how they can utilize this plan to help safeguard the environment.

The bed sheet sign is a small laminated card that is located on the nightstand of each room. The card informs the guest that the hotel customarily changes the bed sheets every day; however, if the guest doesn't feel this is necessary, he or she can leave the card on the bed. This serves as a signal to hotel workers that the bed sheets don't require changing. The Healthy Earth Hotel laundry hangers work similarly. These hangers feature two rungs and an explanatory note asking guests to keep clean towels on the top rung, and hang soiled towels on the bottom rung. This way, hotel personnel don't need to wash both the soiled and the clean towels.

The benefits of this safe and simple program are basically boundless. From the perspective of a hotel proprietor, the Healthy Earth Hotel plan is fiscally brilliant—it will save incredible amounts of money and resources. Hotel managers spend literally millions of dollars each year on the electricity, water, detergents, and washing machines needed to do the laundry. Cutting unnecessary financial burdens, such as laundering linens that aren't soiled, can save a bundle. When hotel managers save money, they can afford to offer more reasonable rates to their guests.

Saving money is just one of many reasons hotel owners and guests alike support the Healthy Earth Hotel plan. Another reason is that the program does not involve any unreasonable sacrifice. Not many people wash towels and bed sheets after a single day's use, so why would hotels be expected to? Guests understand this principle, and will appreciate the hotel's sensitivity to environmental concerns.

The ways hotels can save money and help to protect the environment are almost endless. That's why the Healthy Earth Hotel program offers many other services in addition to the laundry plans described above. Our program also offers plans relating to hotel usage of water, electricity, and gas. For instance, did you know that the average hotel guest spends over ten minutes showering each day? For most showers, that means over eleven gallons of water going down the drain! That's a lot of water. Additionally, long showers add more pollutants into the water supply, such as soap and shampoo. Studies have shown that people can shower thoroughly in under five minutes, and that long showers are just a big waste of energy. The Healthy Earth Hotel program offers wall hangings that summarize these statistics for guests and urge them to take shorter showers. In the long run, this will benefit everyone.

Looking at a broader perspective, there's an even more positive aspect to this plan. The Healthy Earth Hotel program helps to preserve the planet's precious natural resources. All of the materials and energy that hotels use for their laundering either subtract from the world's resources or cause pollution. Reducing the amount of laundry a hotel does is an automatic benefit to the environment. Many people go on vacations in order to enjoy the splendor of nature, whether it's to the Grand Canyon, a balmy beach, or a sparkling rural lake. Since hotels rely on these travelers, shouldn't they help to keep nature beautiful? The answer is a resounding **yes!** The Healthy Earth Hotel should be implemented whenever possible.

To register for the Healthy Earth Hotel program, or to request more information, visit our local office at the Chesterfield Mall or call 555-HEHP.

 Questions

1   The author MOST LIKELY wrote this passage

    A   to inform readers about the Healthy Earth Hotel program.
    B   to convince hotels to consider the Healthy Earth Hotel program.
    C   to persuade readers that the Healthy Earth Hotel program is a good idea.
    D   to offer a comparison of the pros and cons of the Healthy Earth Hotel program.

 Tip

Consider the audience. To whom is the author writing?

2   Which sentence BEST tells about this selection?

A   People travel to escape the stresses of their daily lives, and they expect a clean, comfortable environment when they stay in a hotel.

B   The Healthy Earth Hotel program helps the environment by preserving the earth's precious resources and saving hotel owners money.

C   When hotels start the Healthy Earth Hotel program, they receive a start-up kit with special hangers and signs.

D   The Healthy Earth Hotel program can save millions of dollars normally spent on water, electricity, detergents, and washing machines.

 **Tip**

Choose the answer choice that best summarizes the entire article.

3   According to the author, which is PROBABLY the biggest benefit from the Healthy Earth Program?

A   It saves a lot of money.
B   It pleases hotel guests.
C   It helps the environment.
D   It reduces hotel rates.

 **Tip**

Reread the second-to-last paragraph of the article for a clue.

Now check your answers on the next page.

Passage 1: "Healthy Earth Hotels"

 Answers

1   C   The passage is persuasive. The author is strongly in favor of the Healthy Earth Hotel program. With this in mind, both answer choices B and C might seem correct. However, the author is not writing to hotel owners. He is telling readers—the general public—about the program. Therefore, answer choice C is the best answer.

2   B   The best summary is answer choice B. It correctly summarizes the entire article. The other answer choices give details in the article.

3   C   In the past paragraph of the article, the author says, "there's an even more positive aspect to this plan." This hints that he considers this aspect—the environment—to be the most important.

Passage 2

Read the following poem and answer the questions that follow.

The Lamplighter
by Robert Louis Stevenson

My tea is nearly ready and the sun has left the sky.
It's time to take the window to see Leerie going by;
For every night at teatime and before you take your seat,
With lantern and with ladder he comes posting up the street.

Now Tom would be a driver and Maria go to sea,
And my papa's a banker and as rich as he can be;
But I, when I am stronger and can choose what I'm to do,
O Leerie, I'll go round at night and light the lamps with you!

For we are very lucky, with a lamp before the door,
And Leerie stops to light it as he lights so many more;
And oh! before you hurry by with ladder and with light;
O Leerie, see a little child and nod to him to-night!

 **Questions**

1   Which of the following BEST describes the subject of this poem?

  A   a lamplighter named Leerie
  B   a child who wants to see a lamplighter
  C   a child who likes to look out the window each night
  D   a child who wonders what to be when he grows up

 **Tip**

Choose the answer choice that tells what the entire poem is about. Consider who is speaking in the poem.

2   Which is one theme of this poem?

    A   Children find beauty in simple things.

    B   Children view the world differently than adults.

    C   Some of the best jobs are the most common.

    D   It is fun to watch people do their jobs.

 **Tip**

    Think about the way the child feels about Leerie.

3   Which is the BEST summary of the second stanza of the poem?

    A   The child's friends know what they want to be when they grow up.

    B   The child's father is a banker and has a lot of money.

    C   The child will choose what he wants to do when he is stronger.

    D   The child would like to be a lamplighter when he grows up.

 **Tip**

    Reread the second stanza. What is it mostly about? Eliminate answer choices that give only details.

Now check your answers on the next page.

Passage 2: "The Lamplighter"

 **Answers**

1   B   While the first answer choice might also seem correct, the child is an important part of the poem along with the lamplighter. Answer choice B includes both the child and the lamplighter, so it is the best answer choice.

2   B   Answer choice B is the best answer because the child really likes to watch the lamplighter. He finds this job enjoyable.

3   D   The second stanza of the poem is mostly about what the child wants to do when he grows up: he wants to light the lamps with Leerie.

Passage 3

Read this passage. Think about the author's purpose as you read. When you finish, answer the questions that follow.

Come to Camp Wallabe!

Stifling in the sticky city heat? Seen the same summer sights over and over again? It's time to abandon your apartment, gather up your gear, and head to Camp Wallabe for some summertime fun!

Located along the pristine shores of Haida Lake, Camp Wallabe offers a diverse selection of summer activities that are sure to satisfy every camper. Choose one of two five-week programs offered to campers ages five to sixteen between the months of June and August.

Location
The lakeside grounds of Camp Wallabe are cared for by campers. Teaching children to respect their environment by cleaning up after themselves and by caring for the natural resources around them is an important part of the Camp Wallabe experience.

Participating in various exciting activities on the camp's extensive grounds becomes a reward for the hard work necessary to keep our camp clean.

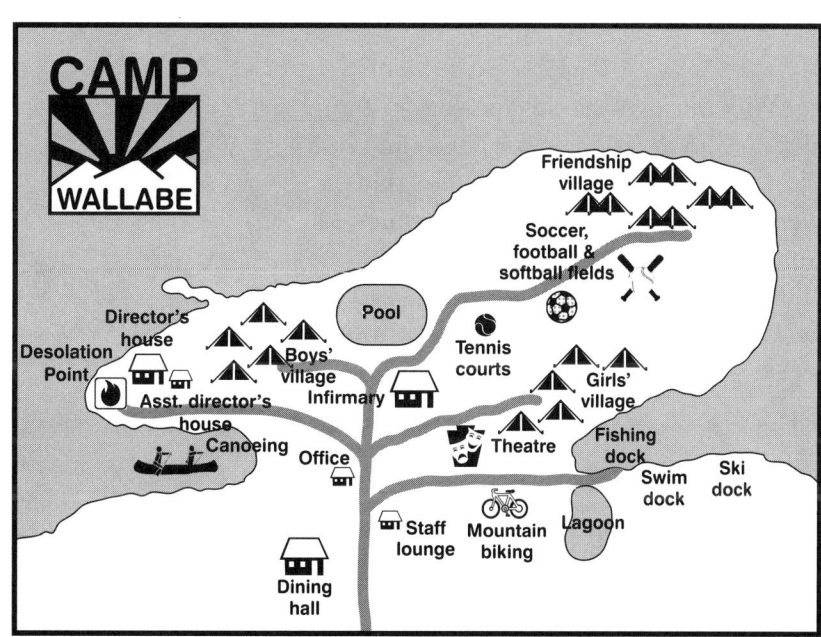

Activities

Camp Wallabe offers more activities for campers than any other facility in the area.

Swimming—Beginning to intermediate swimmers can cool off all summer long in one of the camp's two pools, or in refreshing Haida Lake itself. Our counselors are all certified lifeguards trained in swimming instruction, CPR, and first aid. Campers can participate in swimming instruction or enjoy leisure swims during lifeguard hours. Advanced swimmers may perform in program-long structured aquatic competitions.

Boating/Fishing—Counselor-supervised boating activities include kayaking, canoeing, sailing, and fishing. At the end of each program, campers may participate in a day-long fishing adventure that ends with a fish fry on the beach. Campers may also sign up for kayak and canoe racing in the Wallabe Wars end-of-program camp-wide competition. Campers twelve and older may also take waterskiing lessons.

Athletics—Camp Wallabe has so many athletic activities you may not get to try them all! The camp has multiple facilities for playing soccer, softball, football, lacrosse, volleyball, basketball, field hockey, archery, rock climbing, and more. Campers may choose from several athletic opportunities on a daily basis.

Horseback Riding—With our own stables located on camp grounds, Camp Wallabe is an ideal place for beginning and advanced riders alike. Our stables are run by equestrian experts and house several steeds, as well as a few ponies for younger riders. Campers can go on guided trail and shoreline rides or practice guiding horses around a course, complete with small obstacles. Horses draw students in a large wagon for an end-of-summer hay ride.

Theater Arts—Many campers choose to pursue a serious summer study of the dramatic arts, attending daily acting and performance workshops to improve theatrical skills. In addition to performing weekly skits and musical compositions for campers and staff alike, campers may participate in the production of a full-length play—which also offers many behind-the-scenes opportunities for involvement—to be held at the end of each program. Students who play a musical instrument are welcome to store their instruments in secure camp facilities.

Arts and Crafts—From beaded key chains to sculpture and sofa-size paintings, Camp Wallabe is prepared to accommodate the artist in each camper. Our arts and crafts rooms are spacious and stocked with supplies. Campers may spend daily craft time creating pottery, ceramics, origami, jewelry, and papier mâché designs. Quilting and other textile activities are also favorites of many campers. Those campers who are serious about art may enroll for sketching, sculpture, photography, and oil painting classes, at an extra cost. These classes require that campers dedicate a certain amount of time each day to artistic projects to make sure that projects are completed by the end of the program. Students enrolled in these classes get preference for studio time.

Schedule

Each week, campers may structure their own schedules, with some flex time for trying out unscheduled activities.

    7:30 AM: Rise and shine, campers!
    7:45 AM: Outdoor yoga (optional)
    8:00 to **9:00** AM: Breakfast/cleanup in the mess hall
    9:00 to **10:30** AM: Activity 1 (from weekly schedule)
    10:30 to **11:45** AM: Activity 2 (from weekly schedule)
    12:00 to **1:00** PM: Lunch/cleanup in the mess hall
    1:00 to **3:00** PM: Flex time (choose an activity)
    3:00 to **5:00** PM: Activity 3 (from weekly schedule)
    5:00 to **5:30** PM: Camp-wide cleanup
    5:30 to **6:00** PM: Pre-dinner cabin time
    6:00 to **7:30** PM: Dinner/cleanup in the mess hall
    8:00 to **10:00** PM: Mess Hall Movie Series (wear your PJs!)
    10:15 PM: Lights out.

Costs

Camper costs vary each season. Costs include meals, lodging, and most activities. Additional costs include Camp Wallabe T-shirts, sweatshirts, sweatpants, and shorts (all required). Certain art classes also carry an additional cost.

    Contact the camp for the most recent prices, to sign up for special classes, and to find out more about summers at Camp Wallabe!

 Questions

1   A camper who cleans up around the lake is MOST like

   A   a cat that eats all of the food in its dish.
   B   a waiter who clears away dirty dishes.
   C   a doctor who treats her patients.
   D   a father who does the family's laundry.

 **Tip**

You have to use your own experience and details in the article to answer this question. If you need help, think about the main idea of the article.

2   With which statement would the author MOST likely agree?

   A   Horseback riding is the best activity at Camp Wallabe.
   B   Most campers can find an activity that they like at Camp Wallabe.
   C   Too much money is spent on arts and crafts at Camp Wallabe every year.
   D   Anyone pursuing a theatrical career should attend Camp Wallabe.

 **Tip**

Reread the passage and eliminate incorrect answer choices. Which answer choice is true based on the information in the passage?

3   Which kind of passage did the author write about Camp Wallabe?

   A   an amusing passage that retells true stories about kids at the camp
   B   a persuasive passage that convinces readers to become serious about studying art at the camp
   C   an informative passage that tells readers what the camp is like
   D   a helpful passage that teachers readers about athletic activities at the camp

 **Tip**

Go back and reread the article. What is its purpose?

4   According to the author, a person taking sculpture is MOST LIKELY to

    A   be drawn toward boating.
    B   swim in the lake rather than the pool.
    C   make time to attend campers' skits.
    D   spend flex time in the studio.

Tip

What point the author make when discussing sculpture? How are campers that take sculpture different than other campers?

Now check your answers on the next page.

Passage 3: "Come to Camp Wallabe!"

 **Answers**

1   D   A father who does the family's laundry is cleaning up after himself and others, which is what the campers must do.

2   B   The author's message in the article is that the camp contains so many different activities that every camper will have a good time.

3   C   The purpose of this passage is to let people know what the camp is like. It lists all of the different components of the camp so that readers can learn about it.

4   D   The author states that people who take sculpture are required to dedicate a certain amount of time a day to their artistic products, meaning that they will have less free time than other campers.

Passages 4 and 5

Read the passages in this section and then answer the questions.

Excerpt from *The Prince and the Pauper*
by Mark Twain

London was fifteen hundred years old, and was a great town—for that day. It had a hundred thousand inhabitants—some think double as many. The streets were very narrow, and crooked, and dirty, especially in the part where Tom Canty lived, which was not far from London Bridge. The houses were of wood, with the second story projecting over the first, and the third sticking its elbows out beyond the second. The higher the houses grew, the broader they grew. They were skeletons of strong criss-cross beams, with solid material between, coated with plaster. The beams were painted red or blue or black, according to the owner's taste, and this gave the houses a very picturesque look. The windows were small, glazed with little diamond-shaped panes, and they opened outward, on hinges, like doors.

The house which Tom's father lived in was up a foul little pocket called Offal Court, out of Pudding Lane. It was small, decayed, and rickety, but it was packed full of wretchedly poor families. Canty's tribe occupied a room on the third floor. The mother and father had a sort of bedstead in the corner; but Tom, his grandmother, and his two sisters, Bet and Nan, were not restricted—they had all the floor to themselves, and might sleep where they chose. There were the remains of a blanket or two, and some bundles of ancient and dirty straw, but these could not rightly be called beds, for they were not organised; they were kicked into a general pile, mornings, and selections made from the mass at night, for service. . . .

No, Tom's life went along well enough, especially in summer. He only begged just enough to save himself, for the laws against mendicancy[1] were stringent, and the penalties heavy; so he put in a good deal of his time listening to good Father Andrew's charming old tales and legends about giants and fairies, dwarfs and genii, and enchanted castles, and gorgeous kings and princes. His head grew to be full of these wonderful things, and many a night as he lay in the dark on his scant and offensive straw, tired, hungry, and smarting from a thrashing, he unleashed his imagination and soon forgot his aches and pains in delicious picturings to himself of the charmed life of a petted prince in a regal palace. . . .

He often read the priest's old books and got him to explain and enlarge upon them. His dreamings and readings worked certain changes in him, by-and-by. His dream-people were so fine that he grew to lament his shabby clothing and his dirt, and to wish to be clean and better clad. He went on playing in the mud just the same, and enjoying it, too; but, instead of splashing around in the Thames solely for the fun of it, he began to find an added value in it because of the washings and cleansings it afforded. . . .

By-and-by Tom's reading and dreaming about princely life wrought such a strong effect upon him that he began to ACT the prince, unconsciously. His speech and manners became curiously ceremonious and courtly, to the vast admiration and amusement of his intimates. But Tom's influence among these young people began to grow now, day by day; and in time he came to be looked up to, by them, with a sort of wondering awe, as a superior being. He seemed to know so much! and he could do and say

[1]mendicancy: begging

such marvellous things! and withal, he was so deep and wise! Tom's remarks, and Tom's performances, were reported by the boys to their elders; and these, also, presently began to discuss Tom Canty, and to regard him as a most gifted and extraordinary creature. Full-grown people brought their perplexities to Tom for solution, and were often astonished at the wit and wisdom of his decisions. In fact he was become a hero to all who knew him except his own family—these, only, saw nothing in him.

Excerpt from *My Antonia*
by Willa Cather

When we reached the level and could see the gold tree-tops, I pointed toward them, and Antonia laughed and squeezed my hand as if to tell me how glad she was I had come. We raced off toward Squaw Creek and did not stop until the ground itself stopped—fell away before us so abruptly that the next step would have been out into the tree-tops. We stood panting on the edge of the ravine, looking down at the trees and bushes that grew below us. The wind was so strong that I had to hold my hat on, and the girls' skirts were blown out before them. Antonia seemed to like it; she held her little sister by the hand and chattered away in that language which seemed to me spoken so much more rapidly than mine. She looked at me, her eyes fairly blazing with things she could not say.

"Name? What name?" she asked, touching me on the shoulder. I told her my name, and she repeated it after me and made Yulka say it. She pointed into the gold cottonwood tree behind whose top we stood and said again, "What name?"

We sat down and made a nest in the long red grass. Yulka curled up like a baby rabbit and played with a grasshopper. Antonia pointed up to the sky and questioned me with her glance. I gave her the word, but she was not satisfied and pointed to my eyes. I told her, and she repeated the word, making it sound like "ice." She pointed up to the sky, then to my eyes, then back to the sky, with movements so quick and impulsive that she distracted me, and I had no idea what she wanted. She got up on her knees and wrung her hands. She pointed to her own eyes and shook her head, then to mine and to the sky, nodding violently.

"Oh," I exclaimed, "blue; blue sky."

She clapped her hands and murmured, "Blue sky, blue eyes," as if it amused her. While we snuggled down there out of the wind, she learned a score of words. She was alive, and very eager. We were so deep in the grass that we could see nothing but the blue sky over us and the gold tree in front of us. It was wonderfully pleasant. After Antonia had said the new words over and over, she wanted to give me a little chased silver ring she wore on her middle finger. When she coaxed and insisted, I repulsed her quite sternly. I didn't want her ring, and I felt there was something reckless and extravagant about her wishing to give it away to a boy she had never seen before. No wonder Krajiek got the better of these people, if this was how they behaved.

While we were disputing about the ring, I heard a mournful voice calling, "Antonia, Antonia!" She sprang up like a hare. "Tatinek! Tatinek!" she shouted, and we ran to meet the old man who was coming toward us. Antonia reached him first, took his hand and kissed it. When I came up, he touched my shoulder and looked searchingly down into my face for several seconds. I became somewhat embarrassed, for I was used to being taken for granted by my elders.

We went with Mr. Shimerda back to the dugout, where grandmother was waiting for me. Before I got into the wagon, he took a book out of his pocket, opened it, and showed me a page with two alphabets, one English and the other Bohemian. He placed this book in my grandmother's hands, looked at her entreatingly, and said, with an earnestness which I shall never forget, "Te-e-ach, te-e-ach my Antonia!"

 Questions

Questions 1 through 3 refer to "Excerpt from *The Prince and the Pauper*."

1   Which of the following BEST describes the subject of this passage?

A   a boy dealing with problems caused by location
B   a boy dealing with problems caused by jealousy
C   a boy dealing with problems caused by a lack of care
D   a boy dealing with problems caused by a lack of money

 Tip

This question refers to the central idea of the passage. What is it mainly about?

2   The major change in Tom's mannerisms and behaviors are PROBABLY because

A   Tom resents his station in life when he sees the princely lifestyle.
B   Tom makes fun of the princely lifestyle by imitating it.
C   Tom learns about the princely lifestyle and imitates it.
D   Tom dreams about becoming princely one day.

 Tip

Think about why Tom's behavior changed. What made him change the way he acts?

3   Which is one theme of this story?

   A   Your family should believe in you.

   B   Be careful what you wish for.

   C   Dreams are sometimes confused with reality.

   D   If you tell a lie, you might start to believe it.

 **Tip**

   Reread the story. What caused Tom to start acting extraordinary?

Questions 4 through 6 refer to "Excerpt from *My Antonia.*"

4   Which is the BEST summary of this story?

   A   A man gives a book to a little boy.

   B   A girl and her sister meet a boy.

   C   A girl tries to give a boy her ring.

   D   A boy teaches a girl some English words.

 **Tip**

   Begin by eliminating answer choices that are not true or that present details in the story, but not a summary of the whole story.

5   The narrator refuses Antonia's ring because he

   A   doesn't think boys should wear rings.

   B   thinks her father will be mad if he takes it.

   C   doesn't think she should give her ring to a stranger.

   D   thinks his mother will make him give it back.

 **Tip**

Eliminate answers that the text does not support.

6   Why did Mr. Shimerda show the book to the narrator's grandmother?

    A   He wants her to teach his daughter English.

    B   He wants to show her his alphabet.

    C   He wants her to learn his language.

    D   He wants her to read to him.

 **Tip**

Think of all the things that are said as Mr. Shimerda shows her the book.

Questions 7 and 8 refer to both "Excerpt from *The Prince and the Pauper*" and "Excerpt from *My Antonia*."

7   Which idea is explored in BOTH passages?

    A   not fitting in with your family

    B   traveling to a new place

    C   learning something new

    D   meeting someone for the first time

 **Tip**

Think about whether each idea is covered in BOTH passages.

8   A word that describes BOTH Tom and the narrator in *My Antonia* is

    A   shy

    B   sullen

    C   loud

    D   intelligent

 **Tip**

Think about each character's actions. Which word describes them BOTH?

Now check your answers on the next page.

Passage 4: Excerpts from *The Prince and the Pauper* and *My Antonia*

 Answers

1   D   While some of Tom's problems may be caused by a lack of care, most are caused by his family's poverty. Answer choice D is the best answer.

2   C   Though answer choices A and D might seem right, they do not describe the major change in Tom's mannerisms and behavior during the passage. Choice B is never established in the passage, because Tom never mocked the princely life, but rather, dreamt of living it. Answer choice C is the best answer because it says in the passage that Tom's speech and mannerisms changed subconsciously and he began acting very much like a petted prince.

3   C   This answer choice touches on the central idea of the passage: Tom doesn't realize that he is acting like a prince. He is confusing his dreams with reality.

4   D   The boy teaches some English words to a girl.

5   C   The boy says that it is reckless of the girl to give her ring to someone she's known for such a short time.

6   A   After he shows the book to the narrator's grandmother, he asks her to teach his daughter English.

7   C   Both passages include characters who are learning something new.

8   D   Both Tom and the narrator of *My Antonia* are very intelligent.

Lesson 5: Writing and Revising

Writing Genres

W.GN.08.01 Write a cohesive, narrative piece that includes appropriate conventions to the genre (e.g., historical fiction, science fiction, realistic fiction), and employ literary and plot devices (e.g., narrator credibility, rising and falling action, and/or conflict, transitional language, and imagery).

W.GN.08.02 Write a historical expository piece (e.g., journal, biography, simulated memoir) that includes appropriate organization, illustrations, marginal notes, and/or annotations.

Writing Process

W.PR.08.01 Set a purpose, consider audience, and replicate authors' styles and patterns when writing narrative or informational text.

W.PR.08.02 Apply a variety of prewriting strategies for narrative text (e.g., story maps designed to depict rising and falling actions, roles of minor characters, credibility of narrator) and informational text (e.g., compare/contrast, cause and effect, sequential text patterns).

W.PR.08.03 Experiment with various ways of sequencing information (e.g., ordering arguments, sequencing ideas chronologically or by importance).

W.PR.08.04 Review and revise compositions for coherence and consistency regarding word choice, cause and effect, and style, and they will read their own work from another reader's perspective in the interest of clarity.

W.PR.08.05 Edit their writing, using proofreaders' checklists, both individually and in peer editing group.

Personal Style

P.PS.08.01 Exhibit individual style to enhance the written message (e.g., in narrative text: personification, humor, element of surprise; in informational text: emotional appeal, strong opinion, credible support).

Grammar and Usage

W.GN. 08.01 Use style conventions (e.g., MLA) and a variety of grammatical structures in their writing, including infinitives, gerunds, participial phrases, and dashes or ellipses.

Spelling

WSP.08.01 Students will use correct spelling conventions in the context of their own writing.

You will be asked to write the following essays on MEAP. You will also be asked to answer questions about revising a student essay.

- Constructed-response question (Reading, Part 1B)
- Writing from Knowledge and Experience essay (Writing, Part 4A)
- Question about student essay (Writing, Part 4B)

Before you learn more about each essay on the MEAP, read the following section on essay-writing in general.

Developing Your Essays: The Three Stages of Writing

As you begin to develop your essay, you should follow the three stages of writing: prewriting, drafting, and revising.

Prewriting—RECORD Your Ideas

The main purpose of prewriting is to record your ideas. You can do this by brainstorming what you will be writing about in your essay. Start by jotting down ideas and possible angles for your essay. Think about the positive and negative aspects of a topic. Think about the audience for whom you will be writing and the purpose of your writing. Are you writing to entertain readers with a story, give your opinion, or explain something? Once you have determined your central idea, purpose, and audience, write down some supporting material and organize or outline your ideas into a logical sequence.

Suppose your task is to write an essay in support of or against tearing down an old building.

> At the last city council meeting, a local business owner asked the council members for permission to tear down a historic building on Main Street to build a new clothing store in its place. Council members were divided on the issue. Some argued that the building was built before the Civil War and had too much historic value to be destroyed. Others argued that the old building was nothing more than an eyesore and a safety hazard and that a new store would make the downtown area more attractive.
>
> The mayor decided to postpone voting on the issue until she could hear more details about both sides of the issue. How do you feel about tearing down the historic building?
>
> Write an essay giving your opinion on the issue. Use facts and examples to develop your argument.

How would you begin to prepare an essay on this issue? First, you would take a moment to jot down a few notes about the issue. Why is the old building important? What are the benefits of the new store? You would ask yourself how you feel about the issue. Do you disagree with tearing down the building, or would you rather have a new clothing store? Once you decide on the angle you want to take in your essay, add some details to support your position. You could create a web to help develop your argument and organize your ideas.

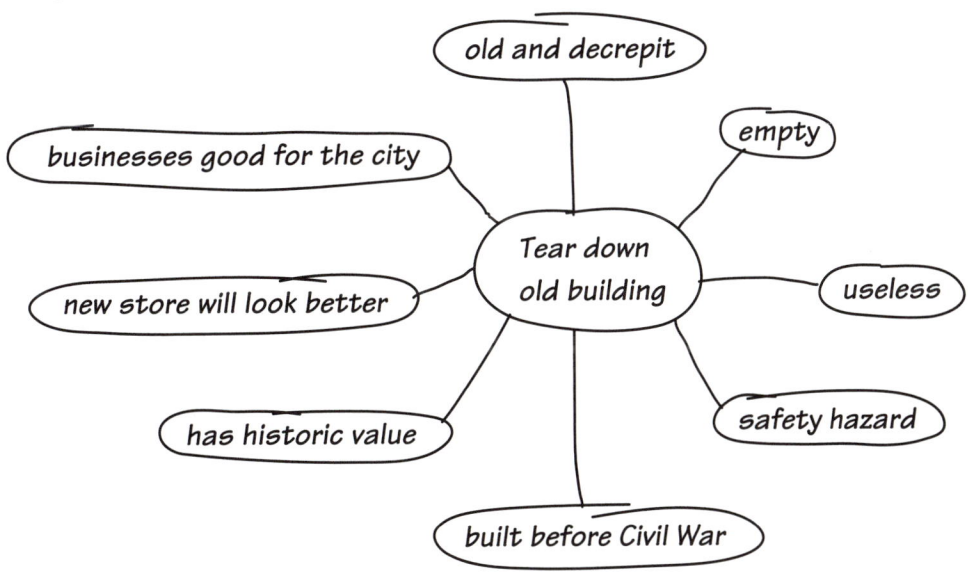

Prewriting— EVALUATE Your Ideas

When you have finished developing your argument, evaluate what you've written. Which ideas will help persuade the reader to share your opinion? Which ideas might weaken your argument? Don't be afraid to eliminate one or more of your ideas.

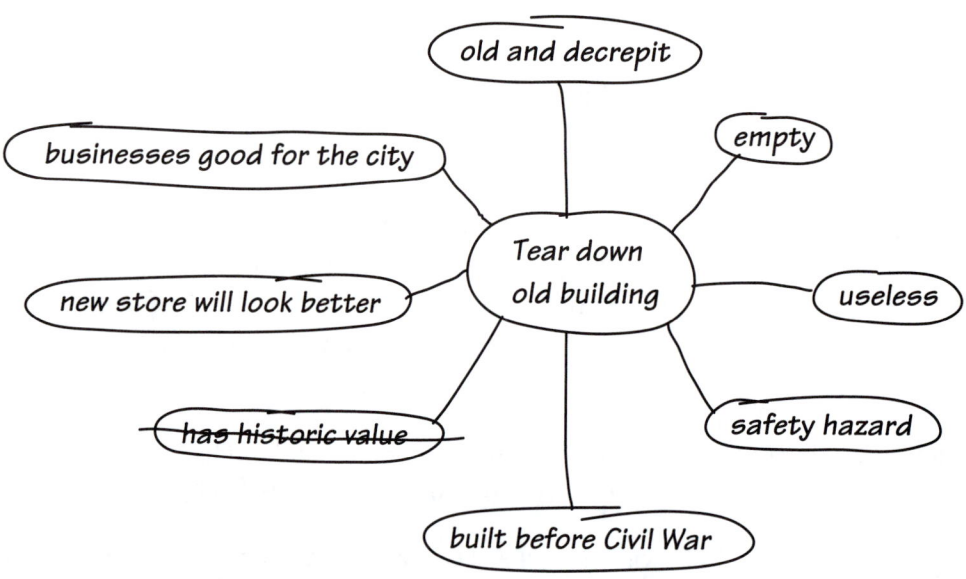

The fact that the building has historic value doesn't support the argument to tear it down. You probably wouldn't want to focus your essay on the historic value of the building if you were trying to convince readers to tear it down.

Prewriting—ORGANIZE Your Ideas

A good essay is organized into three parts:

1. Introduction—An essay should always begin with an introduction. The introduction should give readers a good idea of what to expect in the essay and give them a clue as to why you are writing the essay.

2. Body—The body of the essay is where you present the main ideas of the essay. Your main ideas, or in this example your main arguments, should be clearly explained. State your main ideas or opinions and support them with details.

3. Conclusion—The conclusion should provide a quick summary of your essay and leave the reader with your final word on the issue.

Drafting—Begin Your FIRST DRAFT

In the drafting stage of writing, you will write a rough draft of your work. An important thing to remember when writing your draft is to get your ideas down on paper. In this stage of writing, your writing does not have to be perfect. It is acceptable for the rough draft to have mistakes in grammar, spelling, and punctuation. These mistakes can be changed or fixed later.

Your first draft may look something like this:

> I think that replacing the building with a new clothing store is a grate idea. At the last city council meeting a local business owner asked permishon to tear down the old building on Main St. The business owner wants to build a new clothing store in its place. I think city council should vote for this project.
>
> Right now, the old building is full of broke windows. The doors are missing and bats and rats live their. More than anything, the building is an eye sore. A new building of any kind would look better.
>
> Some city council members have argued that the old building is historic. Because it was built before the civil war. But, they fail to mention that the building has is in disrepare. Building a new store will improve the look of the downtown area. Pieces of broken glass and brick could easily fall to the ground and hurt people on the sidewalk the building is just not safe.
>
> Finally, the old building is empty and useless. Bilding a new store in it's place would bring more people and more money into the city. It's taking up a lot of space and it's not being used for anything.
>
> I would like to ask all city council members to think about how wonderful Main St. could look if the unsafe, useless, eyesore of a building was torn down. And replaced with a brand new store.

Revising and Editing—Preparing the FINAL DRAFT

After you write your rough draft, it's time to begin revising and editing your work. Read your rough draft carefully. Look for mistakes in grammar, spelling, punctuation, and capitalization. Look for sentence fragments. Make sure that you have stated your main idea or that you have provided enough supporting details for readers to determine the central theme. Reword sentences or move entire paragraphs to make your writing flow in a clear, logical order. Add more details to make your writing vibrant and exciting.

Editorial Symbols

When you edit your first draft, you will find it helpful to use editorial symbols. These are marks on the page that show how you want your composition to be improved. The most common editorial symbols are

℘	This is a delete symbol. It tells you what should be removed from the text.
∧	This is an insert symbol. It tells you what should be added to the text.
⊙	This symbol tells you to add a period.
≡	This mark under a letter means that it should be changed to an uppercase letter.
/	This mark through a letter tells you that it should be changed to a lowercase letter.
◡	This symbol means that you should delete a word or space and bring the surrounding letters together.

When you have finished revising your first draft, refer to the Checklist for Revising and Proofreading, on the following pages, to help perfect your essay. Make sure that your essay hits each point listed in the checklist. Then write the final copy of your work in the answer booklet of your test.

CHECKLIST FOR REVISING AND PROOFREADING

DIRECTIONS

Use the following checklists as you revise and proofread your writing. When you are finished revising, you must write your final copy. Then, proofread your final copy to make sure that all of your revisions have been made.

Checklist for Revision

_____ Do I have a clear central idea that connects to the theme?

_____ Do I stay focused on the theme?

_____ Do I support my central idea with important details/examples?

_____ Do I need to take out details/examples that DO NOT support my central idea?

_____ Is my writing organized and complete?

_____ Do I use a variety of words, phrases, and/or sentences?

Checklist for Editing

_____ Have I checked and corrected my spelling to help readers understand my writing?

_____ Have I checked and corrected my punctuation and capitalization to help readers understand my writing?

Checklist for Proofreading

_____ Is everything in my final copy the way that I want it?

The final draft of your essay might look something like this:

At the last city council meeting, a local business owner asked permission to tear down the old building on Main Street and construct a new clothing store in its place. I think that replacing the old, rundown building with a new clothing store is a great idea. I encourage city council to vote in favor of this project.

Some city council members have argued that the old building holds a lot of historic value because it was built before the Civil War. However, they fail to mention that the building has fallen into a state of disrepair. Pieces of broken glass and brick could easily fall to the ground and hurt people on the sidewalk. The building is just not safe.

Replacing the old building with a new store will improve the appearance of the downtown area. Right now, the old building is full of broken windows. The doors are missing and it's home to many bats and rats. More than anything, the building is an eyesore. A new building of any kind would be an improvement.

Finally, the old building is empty and useless. It's taking up a lot of valuable property and not being used for anything. Building a new store in its place would bring more people and more money into the city.

In conclusion, I would like to encourage all city council members to think about how wonderful Main Street could look if an unsafe, useless eyesore of a building were removed and replaced with a brand-new store.

In order to achieve the highest score for your essay, make sure that you use the three stages of writing and the Checklist for Revising and Proofreading. Also, pay attention to the content and organization of your essay, as well as usage, sentence construction, and mechanics.

Content/Organization

As mentioned earlier, your essay should be framed by strong opening and closing ideas. Make sure that you have addressed reasons that your issue is important. Conclude by stating why you feel as you do.

In between the opening and closing of your essay are your main ideas. Make sure that your ideas are clear, and that you have included a variety of main ideas and have not simply stressed the same point multiple times. Your ideas should follow a logical progression, meaning that transition from one main idea to another should not be choppy but instead should flow easily from one idea to the next. Your ideas should also be supported by details, or reasons why you believe your ideas to be true. Also, be sure that your transitions from the introduction to the body to the conclusion are fluid, instead of choppy.

Sentence Construction

Make sure that you follow traditional grammar rules when composing sentences. You should check to make sure that you have placed periods and commas in logical places. Make sure that you vary the length and structure of your sentences. This will help to improve your composition.

Usage

When you revise and edit, make sure that you use correct verb tense and agreement. For example, if you are using past-tense verbs to describe something that happened in the past, then make sure that all the verbs describing this past event are in the past tense. Also, look at your pronouns (*I, you, he, she, it, we, they*) to make sure that you have used them correctly. Examine your essay to make sure you have used words that will engage the reader. If you don't like the look or sound of a certain word in your essay, try to replace it with a better one.

Mechanics

Mechanics are the spelling, capitalization, and punctuation in your essay. You are not allowed to use a dictionary during the test, so try to do your best with spelling and capitalization. Using precise spelling, capitalization, and punctuation will make it easier for people to read and understand your essay.

Constructed-Response Question

The first section of the MEAP Reading test contains a pair of selections. These selections may contain a similar subject matter or theme. In Part 1A, you will answer multiple-choice questions about these selections. In Part 1B, you will answer a constructed-response question. This is an open-ended question, meaning you will write out your answer. The constructed-response question on MEAP is a scenario for which you will write a response. You will have to use your own knowledge and details from BOTH selections in your response.

This is the constructed-response question from the Pretest in the beginning of this book. Turn back to the Pretest Answers for a sample response to this question.

Sample Scenario:

PART 1B: RESPONSE TO THE READING SELECTIONS

DIRECTIONS:

Write a response to the scenario question that is stated in the book below. You may use your own ideas and experiences in your response, but you MUST use examples from BOTH reading selections to earn full credit. You may look back at BOTH reading selections at any time. You will have approximately 25 minutes to complete this part of the test.

You may write down ideas, organize your thoughts, or write a rough draft on the paper provided.

Scenario:

Your teacher has asked the students in your class to write an essay defining love. You are ready to begin writing your essay.

Scenario Question:

What is love? Why is it important to people? Explain your answer, using details from BOTH the excerpt from *Little Women* and the poem "Love Is Not All" to support your answer. Be sure to show how the two reading selections are connected or alike.

Use the Checklist for the Response to the Reading Selections to help you with your response.

CHECKLIST FOR REVISING AND PROOFREADING

DIRECTIONS:

Use the following checklists as you revise and proofread the writing you have done for Part 1. When you are finished revising, you must write your final copy. Then, proofread your final copy to make sure that all of your revisions have been made.

Checklist for Revision

_____ Do I have a clear central idea that connects to the theme?

_____ Do I stay focused on the theme?

_____ Do I support my central idea with important details/examples?

_____ Do I need to take out details/examples that DO NOT support my central idea?

_____ Is my writing organized and complete?

_____ Do I use a variety of words, phrases, and/or sentences?

Checklist for Editing

_____ Have I checked and corrected my spelling to help readers understand my writing?

_____ Have I checked and corrected my punctuation and capitalization to help readers understand my writing?

Checklist for Proofreading

_____ Is everything in my final copy the way that I want it?

Grading

Your answer to the constructed-response question on Part 1B will be graded on the following 6-point rubric:

Michigan Educational Assessment Program (MEAP)
Grade 8

Rubric for the Response to the
Paired Reading Selections

6	The student clearly and effectively chooses key or important ideas from each reading selection to support a position on the question and to make a clear connection between the reading selections. The position and connection are thoroughly developed with appropriate examples and details. There are no misconceptions about the reading selections. There are strong relationships among ideas. Mastery of language use and writing conventions contributes to the effect of the response.
5	The student makes meaningful use of key ideas from each reading selection to support a position on the question and to make a clear connection between the reading selections. The position and connection are well developed with appropriate examples and details. Minor misconceptions may be present. Relationships among ideas are clear. The language is controlled, and occasional lapses in writing conventions are hardly noticeable.
4	The student makes adequate use of ideas from each reading selection to support a position on the question and to make a connection between the reading selections. The position and connection are supported by examples and details. Minor misconceptions may be present. Language use is correct. Lapses in writing conventions are not distracting.
3	The student takes a clear position on the question. The response makes adequate use of ideas from one reading selection **or** partially successful use of ideas from both reading selections to support the position. The position is developed with limited use of examples and details. Misconceptions may indicate only a partial understanding of the reading. Language use is correct but limited. Incomplete mastery over writing conventions may interfere with meaning some of the time.

| 2 | The student takes a clear position on the question. There is partially successful use of ideas from one reading selection **or** minimal use of ideas from both reading selections to support the position. The position is underdeveloped. Major misconceptions may indicate minimal understanding of the reading. Limited mastery over writing conventions may make the writing difficult to understand. |
| 1 | The student takes a position on the question but only makes minimal use of ideas from one reading selection **or** the student attempts to support an unclear position with minimal use of ideas from both reading selections. Ideas are not developed and may be unclear. Major misconceptions may indicate a lack of understanding of the reading. Lack of mastery over writing conventions may make the writing difficult to understand. |

Condition Codes for Unratable Papers (Zeros):

A	Off topic
B	Written in a language other than English or illegible
C	Blank or refused to respond
D	Retells or references the reading selections with no connection to the question
E	Responds to the question with no reference to either of the reading selections

Writing from Knowledge and Experience

The first part of the writing portion of MEAP is called "Writing from Knowledge and Experience." You will be given a theme in this part of the test, and you will write an essay about the theme. You will be allowed to write about this theme in any way that you choose. For example, if your subject is fairness, you might write about an instance in which someone acted fairly or unfairly or you may persuade your reader that a certain law should be passed in the essence of fairness.

The following prompt is an example of one you might see in the Writing from Knowledge and Experience section of the state test. **You will have about 50 minutes to complete your essay.**

Sample Prompt:

WRITING ABOUT THE THEME: FAIRNESS AND JUSTICE

We all expect and want to be treated fairly. And we expect justice of some sort if we are not treated fairly. For example, you might expect a friend to apologize for forgetting to invite you to a birthday party. Write about fairness and justice.

Do **ONLY ONE** of the following:

describe a situation in which you or someone you know has been treated fairly or unfairly

OR

define fairness/unfairness as it has affected you or someone else

OR

explain a way to correct an injustice you or someone you know has experienced

OR

persuade the reader that a situation you know or have experienced is fair or unfair

OR

write about the theme in your own way

You may use examples from real life, from what you read or watch, or from your imagination.

You audience will be interested adults.

Sample Essay:

Our Dress Code Is Not Fair!

Last year, our school board made it mandatory for all students in our public school to wear uniforms. While school board members meant to make life easier for some students by having us all look the same, they have actually made life much more difficult.

For starters, the uniforms selected by the school board are downright uncomfortable. We must wear navy pants and girls can also wear navy skirts. We must wear a belt. For heavier students, like me, the belt is uncomfortable and very unattractive. It simply calls attention to the fact that we are not as thin as some of our classmates. Before the dress code, I could wear longer blouses that were not tucked into my jeans. This concealed my waistline and made me look and feel attractive. Now, I must tuck in my shirt and wear a dark brown belt. This does not make me look or feel good.

We must also wear tan-colored knit shirts that are too warm in the summer and too cold in the winter. The only kind of sweater we are allowed to wear in winter is a button-down cardigan. Kids make fun of anyone wearing this sweater because it looks so bad. So guess what? Most kids don't wear one. We simply shiver instead.

Since the dress code rule was passed, some kids have transferred to private schools. They feel that if public schools will not let them express their individuality, they may as well attend a private school. Many of us do not come from wealthy families. We cannot afford to attend private schools. Paying for our new uniforms is a strain on our families. Since the clothes we are required to wear are so unattractive, students will not wear them outside of school. Because of this, our parents must pay for two sets of clothes: one for in school and one for out of school. This is really expensive!

Our dress code is not helping students—it is hurting them. Please, I urge you, to reconsider this policy and go back to the way things were—the better way! Our dress code is simply unfair.

Grading

Your essay will be graded using the following rubric. An outstanding essay will be given 6 points.

Michigan Educational Assessment Program (MEAP)

Writing from Knowledge and Experience
Grade 8

Holistic Scorepoint Descriptions

6	The writing is exceptionally clear and focused. Ideas and content are thoroughly developed with relevant details and examples where appropriate. The writer's control over organization and the connections between ideas moves the reader smoothly and naturally through the text. The writer shows a mature command of language including precise word choice that results in a compelling piece of writing. Tight control over language use and mastery of writing conventions contribute to the effect of the response.
5	The writing is clear and focused. Ideas and content are well developed with relevant details and examples where appropriate. The writer's control over organization and the connections between ideas effectively moves the reader through the text. The writer shows a command of language including precise word choice. The language is well controlled, and occasional lapses in writing conventions are hardly noticeable.
4	The writing is generally clear and focused. Ideas and content are developed with relevant details and examples where appropriate, although there may be some unevenness. The response is generally coherent, and its organization is functional. The writer's command of language, including word choice, supports meaning. Lapses in writing conventions are not distracting.
3	The writing is somewhat clear and focused. Ideas and content are developed with limited or partially successful use of examples and details. There may be evidence of an organizational structure, but it may be artificial or ineffective. Incomplete mastery over writing conventions and language use may interfere with meaning some of the time. Vocabulary may be basic.
2	The writing is only occasionally clear and focused. Ideas and content are underdeveloped. There may be little evidence of organizational structure. Vocabulary may be limited. Limited control over writing conventions may make the writing difficult to understand.
1	The writing is generally unclear and unfocused. Ideas and content are not developed or connected. There may be no noticeable organizational structure. Lack of control over writing conventions may make the writing difficult to understand.

Condition Codes for Unratable Papers (Zeros):

A	Off topic
B	Written in a language other than English or illegible
C	Blank or refused to respond

Student Writing Sample

In Part 4B of the MEAP, you will have to answer a question about a student essay. You will write out your answer to this question. You will also have to answer five multiple-choice questions about different ways to revise the student essay. **You will have about 30 minutes to write your answer to the question and answer the multiple-choice questions.**

(1)   Dear Editer,
(2)   Last week I went to Newbury Public Park. Last week, I was disgusted to see
(3)   the mounds of dirty aluminum cans, broken coffee cups, tattered newspapers, and
(4)   food rappers that littered the ground at Newbury Public Park. More disturbing than
(5)   the heaps of garbage, however, was the woman who finished her diet soda and
(6)   tossed the can to the ground. I could not believe that this woman was so lazy that
(7)   she threw her trash on the ground, rather than walk fifteen feet to the nearest trash
(8)   receptacal. I have to wonder if this woman throws garbage on the floor of her home,
(9)   why does she think that it's acceptable to do so in a public place?
(10)   Newbury Public Park used to be one of the most attractive parts of our town
(11)   and, now, thanks to people like the diet-soda-can woman, it is becoming a landfill. I
(12)   strongly urge city council to do something about this problem. Otherwise Newbury
(13)   Public Park will soon be Newbury Public Dump.
(14)   Sincerely,
(15)   Jack Chaucer

Sample Peer-Response Question: What could the city council do to stop the littering in Newbury Public Park?

_____

_____

_____

_____

Sample Answer:

There are several things that the city council can do to stop the littering in Newbury Public Park. Members of the council could ask the police force to fine people who litter in the park. They could hire someone to clean up the park and post signs warning people that they could be fined for littering. Publishing photographs of the litter in the local newspaper along with an article about what is being done to stop the littering might also make people aware of the problem and less likely to litter.

Grading

The answer to the Peer-Response Question will be graded using the following 4-point rubric:

Michigan Educational Assessment Program (MEAP)

Writing: Peer Response to a Student Writing Sample
Grade 8

Holistic Scorepoint Descriptions

Here is an explanation of what readers think about as they score your writing.

4	The response clearly and fully addresses the task and demonstrates an understanding of the effective elements of writing that are relevant to the task. Ideas are supported by relevant, specific details from the student writing sample. There may be surface feature errors, but they do not interfere with meaning.
3	The response addresses the task and demonstrates some understanding of the effective elements of writing that are relevant to the task. Ideas are somewhat supported with a mix of general and specific relevant details from the student writing sample. There may be surface feature errors, but they do not interfere with meaning.
2	The response demonstrates limited ability to address the task and may show limited understanding of the effective elements of writing that are relevant to the task. Ideas may be supported with vague and/or partially relevant details from the student writing sample. There may be surface features that partially interfere with meaning.
1	The response demonstrates an attempt to address the task with little, if any, understanding of the effective elements of writing that are relevant to the task. The response may include generalizations about the student writing sample with few, if any, details. There may be surface feature errors that interfere with meaning.

Condition Codes for Unratable Papers (Zeros)

A	Off topic or insufficient
B	Written in a language other than English or illegible
C	Blank or refused to respond
D	Summarizes, revises, and/or copies the student sample, making no connection to the question asked

 Questions

1   In which line does the author uses adjectives to describe the park?

   A   Line 2
   B   Line 3
   C   Line 5
   D   Line 7

 **Tip**

Remember that an adjective describes a person, place, or thing.

2   All of the following are ways to punctuate this sentence EXCEPT:

   A   I have to wonder if this woman throws garbage on the floor of her home why does she think that it's acceptable to do so in a public place?

   B   I have to wonder if this woman throws garbage on the floor of her home. Why does she think that it's acceptable to do so in a public place?

   C   I have to wonder if this woman throws garbage on the floor of her home; why does she think that it's acceptable to do so in a public place?

   D   I have to wonder if this woman throws garbage on the floor of her home, and, if not, why does she think that it's acceptable to do so in a public place?

 **Tip**

Remember that two sentences can only be joined with a conjunction, a semicolon, or a period.

3   What is the correct spelling of "Editer" (line 1)

A   Edditer
B   Editir
C   Editter
D   Editor

Tip

How do you say the word?

4   What is the correct spelling of the word "rapper" (line 4)?

A   Wrap
B   Wrapper
C   Rapier
D   Rappir

Tip

Is this word spelled like it sounds, or is there a silent letter in front of it?

5   What is the correct spelling of the word "receptical" (line 8)?

A   Receptacle
B   Recepticle
C   Recepticel
D   Receptacal

Tip

Look closely at the end of the word.

Lesson 5

 **Answers**

1   B   In line 3, the author says uses the words "dirty aluminum cans, broken coffee cups, tattered newspapers" to describe what is in the park.

2   A   In answer choice A, two independent clauses are run together without correct punctuation.

3   D   The word is pronounced ED-i-ter.

4   B   A rapper is a type of singer, but this sentence calls for the word *wrapper*.

5   A   The word should end in *cle*.

POSTTEST

GENERAL DIRECTIONS:

This posttest is divided into three reading and two writing sections. For the reading sections, read each of the passages and then, on your Answer Sheet, answer the questions that follow.

PART 1—READING

PART 1A: PAIRED READING SELECTIONS

DIRECTIONS:

In Part 1A, you will read two selections and answer the multiple-choice questions that follow each selection. You will then answer some questions that will ask you to think about both of the selections. You may look back at these two selections as often as needed during Part 1.

You may underline, highlight, or write notes in this booklet to help you, but you must mark all of your answers in Part 1A of your **Answer Sheets** on page 217.

You may not use any resource materials (dictionaries, grammar books, spelling books, etc.) for any part of this test.

When you have finished Part 1A, STOP.

WAIT. DO NOT GO ON UNTIL TOLD TO DO SO.

DIRECTIONS: Read Selection 1, a selection about Sojourner Truth. Then answer the questions that follow. **You will have about 40 minutes to complete this section.**

Sojourner Truth: Traveling Preacher

From its beginning, Isabella Baumfree's life was shrouded in hardship. She was denied freedom, separated from her parents, and treated like other people's property. Nobody could have imagined that, years later, she would assume the role of a powerful leader under the name Sojourner Truth.

In 1797, before slavery was abolished in the northern states of America, Baumfree was born into slavery in New York. Since her parents were slaves, she was forced to inherit that horrible destiny. As a young woman she was taken from her family and auctioned to different owners.

One of her owners arranged for her to marry another slave, named Thomas, and together they had five children. These children were in turn considered slaves, and most were taken away and sold just as Isabella and her siblings had been. Isabella could no longer tolerate the suffering. She snatched up her smallest child, abandoned Thomas, and escaped her enslavement.

I Sell the Shadow to Support the Substance.
SOJOURNER TRUTH.

Eventually, the state of New York emancipated its slaves. Isabella, Thomas, and all of their children were finally free. However, that announcement did little to alleviate their troubles. To earn a living, Isabella had to become a house servant for the Van Wagenen family. She even used their name temporarily. During her employment with the Wagenens, Isabella discovered that one of her previous owners had illegally sold her son to a slaveholder in Alabama. She recognized the crime that had been committed and courageously chose to bring the case to court. The court ruled in her favor and forced the crooked slave owners to release her son.

This was Isabella's first taste of empowerment. She'd defended her cause successfully. Now she was determined to spread that power to other black people and women who, at the time, had very little power to speak of.

Isabella was an ideal leader. She was about six feet tall, an exceptional height for a man or woman in her day. She also had developed muscular arms and shoulders from hundreds of days spent plowing fields. Her voice was remarkably deep, but she tempered it with her beautiful singing. Although she was illiterate, she was intelligent, humorous, and well-spoken.

She found additional motivation in religion, joining various communes and movements and studying their beliefs. In 1843, she renamed herself Sojourner Truth. A sojourner is a temporary resident; with her new name, she announced her new role as a traveling preacher.

GO ON TO THE NEXT PAGE

Soon, Sojourner Truth had become famous for her speeches on women's rights and anti-slavery topics. Her most well-known speech is remembered for its dominant challenge: "Ain't I a Woman?" She delivered this speech in 1851 at the Women's Convention in Akron, Ohio. In it, she linked the plight of black people in the south with that of white women in the north and called for change. This is part of what Sojourner told her audience:

"That man over there says that women need to be helped into carriages, and lifted over ditches, and to have the best place everywhere. Nobody ever helps me into carriages, or over mud puddles, or gives me any best place—and ain't I a woman? Look at me. Look at my arm! I have plowed and planted, and gathered into barns, and no man could head me—and ain't I a woman? I could work as much and eat as much as a man, when I could get it, and bear the lash as well—and ain't I a woman? I have borne thirteen children, and seen most all sold off to slavery, and when I cried out with my mother's grief, none but Jesus heard me—and ain't I a woman?"

While Sojourner was extremely intelligent, she always expressed herself in a straightforward manner. She concluded her legendary speech by simply saying, "Obliged to you for hearing me, and now old Sojourner ain't got nothing more to say."

The start of the Civil War in 1861 brought many of Sojourner's major concerns to the forefront. However, while the slavery question was being settled on battlefields across the nation, she was planning for life after the war.

By the time Sojourner Truth died in 1883, her contributions had given Americans many new ideas for the future. She was definitely a person ahead of her time. Sojourner spoke out for women's suffrage (voting rights) approximately 70 years before women gained the legal right to vote. She supported the temperance movement, which urged the government to put restrictions on alcohol, at least 50 years before the government decided to attempt it. She even personally challenged Abraham Lincoln to end segregation on street cars in 1864—which was 91 years before Rosa Parks sparked the modern civil rights movement by refusing to give up her seat on a city bus to a white person.

Questions 1–8

1   The author's MAIN purpose in writing was

    A   to describe the history of slavery in America.

    B   to tell readers about an important event in history.

    C   to inform readers about the life of a brave woman.

    D   to convince people to fight for what they believe in.

GO ON TO THE NEXT PAGE

2   This article could BEST be used as a source for a student research project on

    A   history's greatest speeches.

    B   the Civil War.

    C   great civil rights leaders.

    D   the women's suffrage movement.

3   When the narrator said that Isabella enjoyed her "first taste of empowerment," she PROBABLY means

    A   Isabella felt powerful for the first time.

    B   Isabella discovered what it was like to be free.

    C   Isabella realized she was meant to be in charge.

    D   Isabella realized things would soon change.

4   A sojourner would MOST LIKELY

    A   live with others.

    B   live in a large home.

    C   take a vow of silence.

    D   visit people in a town.

5   In her speech, Sojourner repeated the question "Ain't I a woman?" because she

    A   wanted to show that she was stronger than a man.

    B   hoped to shock her audience.

    C   hoped her audience would pity her.

    D   wanted to make her point very clearly.

GO ON TO THE NEXT PAGE

6 After Isabella Baumfree changed her name to Sojourner Truth, she decided to

A become a traveling preacher.

B abandon her husband and escape from slavery.

C win back custody of her son.

D become a housekeeper for the van Wagenens.

7 Sojourner Truth was DIFFERENT from most women of her time because

A she was taller.

B she could not speak.

C she could not travel.

D she had many children.

8 Sojourner's choice of words in her speech creates a mood of

A regret.

B sorrow.

C anxiety.

D inspiration.

GO ON TO THE NEXT PAGE

DIRECTIONS: Read Selection 2, a selection about the first emancipation. Then answer the questions that follow.

The First Emancipation

In January 1863, during the Civil War, President Abraham Lincoln delivered the Emancipation Proclamation. This speech officially declared the end of slavery in America. The southern states, which considered themselves a separate nation, refused to listen to Lincoln's words. However, when the war ended and the United States of America was restored, Lincoln's message of freedom applied to every state. Finally the ancient institution of slavery was demolished in America.

For his heroic achievement, Lincoln has been called "The Great Emancipator." However, many people don't realize that there were emancipations in America long before 1865.

As far back as the 1600s, when Europeans began to colonize America, slavery was considered normal. Many countries participated in the international slave trade, and thousands of slaves were brought to the New World. At first, many slaves were treated more as indentured servants. This meant that they were forced to do labor, but were given freedom after a certain period of time. Later, slavery was officially legalized by the colonies. Then, these servants were considered true slaves, property of their masters.

In the 1700s, slavery was common in the North. New England, the northeastern region of the country, was actually the center of the American slave trade. Thousands of slaves were forced to work in farms, docks, and shipbuilding yards. However, as the end of that century drew near, slavery in the North was shaken by the American Revolution.

When America's Founding Fathers began struggling to free the nation from the tyranny of the British, they realized a great irony. American Nathaniel Niles summarized the problem when he said: "Let us either cease to enslave our fellow men, or else let us cease to complain of those that would enslave us." How could a nation proclaiming all people to be equal continue to allow slavery? During the course of the Revolutionary War, slavery in the North slowly but surely collapsed.

There are many reasons why slavery was abandoned in the North around the time of the war. One reason was that Britain was making money from the international slave trade. By refusing to purchase slaves, Americans were keeping money from the British. There was also growing pressure from religious groups, most notably the Quakers, who condemned slavery and insisted that it be ceased.

The final, killing blow to slavery in the North came during the battles of the Revolution. During the raging conflicts in the northern colonies, both the American and the British armies competed for the support of slaves. Many American colonies declared that any slave who would fight the British would be made free; the British in turn offered freedom to slaves who fought the Americans. Thousands of slaves participated in the Revolutionary War, on both sides, and were freed. Thousands more fled from their owners during the chaotic conflict.

GO ON TO THE NEXT PAGE

Between 1777 and 1804, all of the colonies in the North finally abandoned the terrible institution of slavery. It would take nearly another hundred years, and another bloody war, to end slavery across all of America.

TIMELINE:

1776 – The Quakers in England and Pennsylvania require that members of their church free their slaves.
1777 – Slavery is banned in Vermont.
1780 – Massachusetts declares all men free and equal, including former slaves.
1780 – Pennsylvania passes a law to free slaves gradually.
1784 – Connecticut and Rhode Island choose to free slaves gradually.
1799 – New York follows Pennsylvania, Connecticut, and Rhode Island.
1807 – Federal law makes it illegal for Americans to participate in the slave trade.

Questions 9–16

9 This article is MOSTLY about

A the causes of the Civil War in the United States.

B Abraham Lincoln's Emancipation Proclamation.

C slavery in the North before the American Revolution.

D freeing America from British control.

10 Lincoln is called the "Great Emancipator" because he

A gave a speech declaring the end of slavery in the United States.

B fought to free the United States from British control.

C released slaves in the North during the American Revolution.

D learned about freedom from America's Founding Fathers.

GO ON TO THE NEXT PAGE

11 The author organizes the article by

    A   comparing and contrasting slavery in the North and South.

    B   presenting a problem and providing solutions to fix it.

    C   explaining the events that led up to the American Revolution.

    D   describing the history of slavery in chronological order.

12 The author's purpose in writing this article was PROBABLY

    A   to persuade readers to learn more about the history of slavery.

    B   to inform readers about slavery in the northern states.

    C   to give readers a detailed description of the life of a slave.

    D   to describe to readers a number of anti-slavery laws.

13 The MAIN reason that many slaves fought during the American Revolution was because

    A   it was the best way for them to earn respect.

    B   being a soldier was one of the few jobs open to them.

    C   they were guaranteed freedom by American and British troops.

    D   President Lincoln promised their freedom if they fought.

14 How did slavery in the northern states change between the 1600s and the 1700s?

    A   Slavery was made legal and slaves became the property of their masters.

    B   New England was no longer at the center of the slave trade.

    C   Most slaves were released after serving their masters for a certain period of time.

    D   By the 1700s, slaves in northern states were given the option of buying their freedom.

GO ON TO THE NEXT PAGE

15 Which sentence BEST summarizes paragraph 5?

  A The Founding Fathers of the United States wanted freedom from British control.

  B American leaders realized that slaves wanted freedom just like they did.

  C Slavery decreased when Americans stopped buying slaves from the British.

  D Many slaves found freedom when they agreed to fight in the Revolutionary War.

16 Which sentence BEST supports the idea that the American Revolution was beneficial to slaves in the northern states?

  A "As far back as the 1600s, when Europeans began to colonize America, slavery was considered normal."

  B "Thousands of slaves participated in the Revolutionary War, on both sides, and were freed."

  C "By refusing to purchase slaves, Americans were keeping money from the British."

  D "New England, the northeastern region of the country, was actually the center of the American slave trade."

GO ON TO THE NEXT PAGE

CROSS-TEXT QUESTIONS

DIRECTIONS: Questions 17 through 21 ask about *both* of the selections you read. Choose the *best* answer for each question. You may look back at the two selections as often as needed.

17 With which of the following would BOTH authors PROBABLY agree?

   A   Slavery in the North was far worse than slavery in the South.

   B   Slaves were forced to work hard and faced many hardships in their lives.

   C   It is possible to help others by sharing your personal story with them.

   D   The onset of the American Revolution began a decrease in slavery in the United States.

18 Which of the following do BOTH passages include?

   A   information about the history of slavery in the United States

   B   information about the history of women's rights in the United States

   C   a description of the end of slavery in the northern states

   D   a description of the life of a brave person who spoke out against slavery

19 Which idea is explored in BOTH passages?

   A   how women fought for the right to vote

   B   how America changed after the American Revolution

   C   how great speakers of the past changed history

   D   how people found freedom from slavery

GO ON TO THE NEXT PAGE

20 How were Sojourner Truth and Abraham Lincoln SIMILAR?

   A   Both were born in New York.

   B   Both believe that women should have the right to vote.

   C   Both are remembered for speaking out against slavery.

   D   Both were taller than average men and women of the time.

21 Based on the two selections, which statement is true?

   A   The only way to make changes in the world is to spark a revolution.

   B   All people have a right to be free and to be treated equally.

   C   Traveling around the country giving speeches is a good way to reach many people.

   D   Only those with power have the ability to make a difference in the world.

STOP

PART 1B: RESPONSE TO THE PAIRED READING SELECTIONS

DIRECTIONS:

Write a response to the scenario question that is stated below. You may use your own ideas and experiences in your response, but you MUST use examples from BOTH reading selections to earn full credit. You may look back at BOTH reading selections at any time. **You will have approximately 25 minutes to complete this part of the test**.

You may write down ideas, organize your thoughts, or write a rough draft in the "Notes/ Planning" section on pages 188 and 189.

Please write your response on pages 218 and 219 of your **Answer Sheets**.

---

22 **One of the United States' most beloved features is freedom. Citizens cite their freedom of speech or their freedom of religion to defend their actions. Your new president wants to give absolute freedom to people within the United States. This means that people will be able to do whatever they want and will not have to obey laws and rules.**

What sort of problems would arise from absolute freedom? Explain your answer using details from BOTH reading selections to support your answer. Be sure to show how the two reading selections are connected or alike.

Use the Checklist on the next page to help you with your response.

WAIT. DO NOT GO ON UNTIL TOLD TO DO SO.

PART 1B: CHECKLIST FOR REVISING AND PROOFREADING

DIRECTIONS:

Use the following checklists as you revise and proofread the writing you have done for Part 1. When you are finished revising, you must write your final copy. Then, proofread your final copy to make sure that all of your revisions have been made.

CHECKLIST FOR REVISION:

_____ Do I have a clear central idea that connects to the theme?

_____ Do I stay focused on the theme?

_____ Do I support my central idea with important details/examples?

_____ Do I need to take out details/examples that DO NOT support my central idea?

_____ Is my writing organized and complete?

_____ Do I use a variety of words, phrases, and/or sentences?

CHECKLIST FOR EDITING:

_____ Have I checked and corrected my spelling to help readers understand my writing?

_____ Have I checked and corrected my punctuation and capitalization to help readers understand my writing?

CHECKLIST FOR PROOFREADING:

_____ Is everything in my final copy the way that I want it?

GO ON TO THE NEXT PAGE

NOTES/PLANNING

NOTES/PLANNING

PART 2—READING

INDEPENDENT READING SELECTIONS

DIRECTIONS:

In Part 2, you will read a selection and answer the multiple-choice questions that follow. You may look back at the selection as often as needed in Part 2.

You may underline, highlight, or write notes in this booklet to help you, but you must mark all of your answers in Part 2 of your **Answer Sheets** on page 220.

You may not use any resource materials (dictionaries, grammar books, spelling books, etc.) for any part of this test.

When you have finished Part 2, STOP.

WAIT. DO NOT GO ON UNTIL TOLD TO DO SO.

DIRECTIONS: Read Selection 1, an excerpt from *Treasure Island*. Then answer the questions that follow. Then read Selection 2 and answer the questions that follow. **You will have about 30 minutes to complete this section.**

An Excerpt from *Treasure Island*
by Robert Louis Stevenson

I was now, it seemed, cut off upon both sides; behind me the murderers, before me this lurking nondescript. And immediately I began to prefer the dangers that I knew to those I knew not. Silver himself appeared less terrible in contrast with this creature of the woods, and I turned on my heel, and looking sharply behind me over my shoulder, began to retrace my steps in the direction of the boats.

Instantly the figure reappeared, and making a wide circuit, began to head me off. I was tired, at any rate; but had I been as fresh as when I rose, I could see it was in vain for me to contend in speed with such an adversary. From trunk to trunk the creature flitted like a deer, running manlike on two legs, but unlike any man that I had ever seen, stooping almost double as it ran. Yet a man it was, I could no longer be in doubt about that.

I began to recall what I had heard of cannibals. I was within an ace of calling for help. But the mere fact that he was a man, however wild, had somewhat reassured me, and my fear of Silver began to revive in proportion. I stood still, therefore, and cast about for some method of escape; and as I was so thinking, the recollection of my pistol flashed into my mind. As soon as I remembered I was not defenseless, courage glowed again in my heart and I set my face resolutely for this man of the island and walked briskly towards him.

He was concealed by this time behind another tree trunk; but he must have been watching me closely, for as soon as I began to move in his direction he reappeared and took a step to meet me. Then he hesitated, drew back, came forward again, and at last, to my wonder and confusion, threw himself on his knees and held out his clasped hands in supplication.

At that I once more stopped.

"Who are you?" I asked.

"Ben Gunn," he answered, and his voice sounded hoarse and awkward, like a rusty lock. "I'm poor Ben Gunn, I am; and I haven't spoke with a Christian these three years."

I could now see that he was a white man like myself and that his features were even pleasing. His skin, wherever it was exposed, was burnt by the sun; even his lips were black, and his fair eyes looked quite startling in so dark a face. Of all the beggar-men that I had seen or fancied, he was the chief for raggedness. He was clothed with tatters of old ship's canvas and old sea-cloth, and this extraordinary patchwork was all held together by a system of the most various and incongruous fastenings, brass buttons, bits of stick, and loops of tarry gaskin. About his waist he wore an old brass-buckled leather belt, which was the one thing solid in his whole accoutrement.

GO ON TO THE NEXT PAGE

"Three years!" I cried. "Were you shipwrecked?"

"Nay, mate," said he; "marooned."

I had heard the word, and I knew it stood for a horrible kind of punishment common enough among the buccaneers, in which the offender is put ashore with a little powder and shot and left behind on some desolate and distant island.

"Marooned three years agone," he continued, "and lived on goats since then, and berries, and oysters. Wherever a man is, says I, a man can do for himself. But, mate, my heart is sore for Christian diet. You mightn't happen to have a piece of cheese about you, now? No? Well, many's the long night I've dreamed of cheese—toasted, mostly—and woke up again, and here I were."

"If ever I can get aboard again," said I, "you shall have cheese by the stone."

All this time he had been feeling the stuff of my jacket, smoothing my hands, looking at my boots, and generally, in the intervals of his speech, showing a childish pleasure in the presence of a fellow creature. But at my last words he perked up into a kind of startled slyness.

"If ever you can get aboard again, says you?" he repeated. "Why, now, who's to hinder you?"

"Not you, I know," was my reply.

"And right you was," he cried. "Now you—what do you call yourself, mate?"

"Jim," I told him.

"Jim, Jim," says he, quite pleased apparently. "Well, now, Jim, I've lived that rough as you'd be ashamed to hear of. Now, for instance, you wouldn't think I had had a pious mother—to look at me?" he asked.

"Why, no, not in particular," I answered.

"Ah, well," said he, "but I had—remarkable pious. And I was a civil, pious boy. . . . And here's what it come to, Jim, and it begun with chuck-farthen on the blessed grave-stones! That's what it begun with, but it went further'n that; and so my mother told me, and predicked the whole, she did, the pious woman! But it were Providence that put me here. I've thought it all out in this here lonely island, and I'm back on piety. You don't catch me tasting rum so much, but just a thimbleful for luck, of course, the first chance I have. I'm bound I'll be good, and I see the way to. And, Jim"—looking all round him and lowering his voice to a whisper—"I'm rich."

I now felt sure that the poor fellow had gone crazy in his solitude, and I suppose I must have shown the feeling in my face, for he repeated the statement hotly: "Rich! Rich! I says.

GO ON TO THE NEXT PAGE

And I'll tell you what: I'll make a man of you, Jim. Ah, Jim, you'll bless your stars, you will, you was the first that found me!"

Questions 23–30

23 The following sentences appear in the story.

"Three years!" I cried. "Were you shipwrecked?"

"Nay, mate," said he; "marooned."

Based on how the word is used in the story, which of the following BEST describes the meaning of "marooned"?

   A   rescued

   B   thrown

   C   deserted

   D   inhabited

24 Which of the following BEST describes Jim's feelings in the excerpt?

   A   scared, then curious

   B   calm, then anxious

   C   angry, then nervous

   D   trusting, then jealous

25 Jim suddenly regains his courage because

   A   he realizes Ben will not hurt him.

   B   he remembers that he has a pistol.

   C   he has found a way to escape from Ben.

   D   he sees that Ben is a man and not an animal.

GO ON TO THE NEXT PAGE

26 Why might the author make Ben appear wild and animal-like?

    A   to show the reader that Ben has been on the island for a long time

    B   to use Ben as a contrast to the civilized world

    C   to show the reader that Ben has become friends with the island's animals

    D   to show the reader how Ben survived on the island

27 The following sentences appear in paragraph 12 of the excerpt:

Wherever a man is, says I, a man can do for himself. But, mate, my heart is sore for Christian diet.

Based on how the word is used in the excerpt, which of the following BEST describes the meaning of the word "sore"?

    A   painful

    B   wound

    C   sting

    D   aching

28 Which of the following narrative genres is NOT a category that this excerpt could be placed in?

    A   historical fiction

    B   adventure

    C   drama

    D   biography

GO ON TO THE NEXT PAGE

29 In the excerpt Jim compares

A Ben's situation with his own.

B Ben's speech with his own.

C the island with civilization.

D his fear of Silver with his fear of Ben.

30 At the end of the excerpt, Jim feels that Ben has gone crazy because

A he has spent too much time with animals.

B he hasn't had a proper meal in years.

C he has been all alone on the island.

D he hasn't spoken English in three years.

GO ON TO THE NEXT PAGE

DIRECTIONS: Read Selection 2, "The Royal Cemetery at Ur." Then answer the questions that follow.

The Royal Cemetery at Ur

The Discovery

In the 1920s, a team of archaeologists led by Sir Leonard Woolley made an amazing discovery. They excavated an ancient cemetery in what was once Mesopotamia, an area between the Tigris and Euphrates Rivers, most of which is now modern-day Iraq, Kuwait, and Saudi Arabia. The cemetery was located in the ancient Sumerian city-state Ur, which existed over 3,000 years ago. The Sumerians used the cemetery for over five hundred years and it contained about 1,800 bodies and many ancient artifacts. Archaeologists have learned a great deal about the Sumerians and life in ancient times from studying the contents of burial tombs at Ur.

The Sumerians

Most people buried in the Royal Cemetery at Ur were common citizens whose funeral rites consisted of merely wrapping their bodies in a reed mat before burial. About sixteen bodies were buried in "royal tombs," large elaborate underground structures with several rooms called chambers. Royal families—kings, queens, and their families—were buried in these tombs. The Sumerians closely intertwined politics and religion and they considered these individuals to be of great importance. Each Sumerian city-state was ruled by a king, who was also a priest. The Sumerians believed that everything around them was controlled by a god. They believed that the sun, moon, and stars were gods. They also believed that their kings were gods and that they were put on the earth to serve them. They built magnificent burial chambers for their kings and queens because they thought this would please them.

The Sumerians believed that kings and queens could take things with them on their journey to the afterlife. They filled royal tombs with everything they thought people would need on this journey, including clothes, jewelry, riches, weapons—and even people. It was not uncommon for Sumerian citizens to sacrifice themselves because they believed this would allow them to accompany their king in the afterlife, where they could continue to serve him.

Queen Puabi's Tomb

Perhaps the most amazing discovery in the Cemetery at Ur was Queen Puabi's tomb. The queen's tomb was especially valuable because it was discovered intact, meaning it had not been disturbed since the Sumerians closed it thousands of years ago. While Queen Puabi's tomb was built on top of another burial chamber, probably the king's, not much is known about the king, however, because his tomb was looted, or robbed, many years ago, probably when the queen was buried.

Queen Puabi's tomb was extraordinary and demonstrated the Sumerian's advanced skills in architectural design. Her body was laid to rest on a table in the middle of an arched cham-

GO ON TO THE NEXT PAGE

ber in the center of what archaeologists refer to as a death pit. The pit and her burial chamber were filled with exquisite ancient artifacts. Queen Puabi was adorned with an incredible headdress made of gold leaves, ribbons, and strands of beads made from rare stones. She wore a cylindrical seal around her neck bearing her name. Her name was carved into the cylinder using cuneiform, the world's first written language, which was invented by the Sumerians. The queen's body was covered in a beaded cape made from precious metals and stone. The cape stretched from her shoulders to her waist. Beautiful rings were carefully placed on each of her fingers.

Members of the queen's "burial party" were discovered in the death pit. Members of this burial party apparently accompanied the queen into her tomb. Each member of the party dressed formally for the special occasion and enjoyed an enormous feast prior to joining Queen Puabi. The burial party included more than a dozen attendants or servants, five armed men, a wooden sled, and a pair of oxen. Four grooms were buried with the oxen, possibly to care for the oxen in the afterlife.

What happened to the members of the burial party to cause their demise? No one is sure, since they died thousands of years ago. However, Sir Woolley and his teams discovered a gold cup near each of their bodies. They suspect that the attendants probably drank poison so they could go to sleep forever along their queen, who may or may not have been already dead.

Questions 31–38

31 This article is MOSTLY about

    A   Sir Woolley.

    B   Queen Puabi.

    C   the Royal Cemetery at Ur.

    D   the riches discovered in Queen Puabi's tomb.

GO ON TO THE NEXT PAGE

32 Why did the Sumerians build elaborate burial tombs for the kings?

  A   They believed their kings were gods.

  B   They hoped to serve their kings in the afterlife.

  C   They thought their kings would return one day.

  D   They wanted to protect their kings from thieves.

33 The following sentence appears in the last paragraph of the article.

  What happened to the members of the burial party to cause their demise?

  Based on how the word is used in the article, which of the following BEST describes the meaning of "demise"?

  A   disease

  B   failure

  C   death

  D   pain

34 By writing this article, the author is trying to

  A   explain what it was like to be a Sumerian.

  B   teach readers about Queen Puabi's tomb.

  C   inform readers about an ancient discovery.

  D   convince readers of the Sumerians' greatness.

35 The information in this article would be BEST used in

  A   an autobiography about Sir Leonard Woolley.

  B   a brochure for a class on ancient civilizations.

  C   an essay encouraging people to visit their local museums.

  D   a research report on the Sumerians.

GO ON TO THE NEXT PAGE

36 The author uses the picture to show the reader

    A   how fashion has changed.

    B   the riches found in Queen Puabi's tomb.

    C   how common Sumerians dressed.

    D   what people looked like during Queen Puabi's time.

37 Sir Woolley and his team suspect that Queen Puabi's burial party died of

    A   starvation.

    B   disease.

    C   poisoning.

    D   suffocation.

38 What is the MAIN idea of the second paragraph of the article?

    A   Queen Puabi's tomb

    B   common Sumerian citizens

    C   the royal tombs in the cemetery

    D   the importance of Sumerian kings

STOP

PART 3—READING

LINKED SELECTION

DIRECTIONS: Read this selection and answer the questions that follow. **You will have 30–50 minutes to complete this section.**

Ouch!

Enrico slipped off his jacket and tossed it on the lawn beside him. The bright sun had warmed the chilly morning air. Enrico picked up a marigold and began removing the dirt around its roots before placing it into one of the shallow holes he and his grandfather had dug. "No, no, Enrico, don't do that," his grandfather gently chided. "It's better to leave the root ball intact. If you break off the soil, you might damage the roots and kill the plant."

Enrico smiled and placed the flower in the hole with the roots and surrounding soil intact. His grandfather, his *abuelo*, was a man of great experience and wisdom. He had taught Enrico many things throughout the years. He had taught him the importance of using good-quality lumber while they built a garden shed together in the backyard. Under his grandfather's guidance, Enrico had learned how to make the world's greatest tortillas using peppers so hot they burned your lips. Most of the novels Enrico devoured at night were his grandfather's picks. Grandfather often recommended a new read for Enrico, usually one that was obscure but wonderful, the kind of book you might find at a yard sale rather than a bestseller list.

Now Enrico wanted his grandfather to teach him how to be a better writer. Even though he had learned English as a second language, Grandfather had become adept enough to publish several history books and had worked for many years as a news reporter for their local paper. "I want to be a writer," Enrico announced as he gently pushed the soil around the marigold so it stood straight in its new home. "I've actually written several short stories. I was hoping you would read them and tell me what you think—and not just tell me that they're good. I want your guidance, so I can improve my writing abilities and become a published writer."

Grandfather shook his head. "Ah, I don't know, Enrico. It isn't easy having your writing critiqued, and I wouldn't want to chance hurting your feelings. You're a bright young man. If you choose to become a writer, you will be a great one, with or without my help."

GO ON TO THE NEXT PAGE

Despite his grandfather's warnings, however, Enrico finally persuaded him to critique one of his short stories. Enrico selected his best piece: a tale about a boy who was the smallest and worst player on his basketball team. The boy, Miguel, was often teased by the other players for his lack of height and skill. Enrico had revised the story several times until he was certain that his command of the English language was at its best. He printed a copy of the story and left it on his grandfather's kitchen table with a note on top that said, "Tell the truth, Grandfather. I can take it, really I can. And I greatly appreciate your help."

After school the next day Enrico rushed to his grandfather's house. "Did you read it?" he asked as his grandfather rubbed a wet dish with a towel and placed it on a shelf.

"Of course I read it, son," Grandfather said and grinned, "and I think you're a very good writer."

Feeling frustrated, Enrico sighed and plopped in a kitchen chair. This was not what he wanted to hear. "Could you give me more than that?" he asked as he ran his fingers through his hair. "Could you tell me what is good about the story and what is bad? Please?"

Grandfather pulled out a chair and sat down beside Enrico. He picked up the printout of Enrico's story. "You need to show more and tell less, for starters," Grandfather advised. "For example, don't tell the reader that Miguel had a sad look on his face. What about Miguel's face looked sad? Describe his face and let your readers draw their own conclusion." Enrico nodded. "And your plot is so predictable that I knew Miguel was going to score the winning basket after reading only the first paragraph. Why not have him miss the shot, but impress his teammates with his newly acquired skill?"

Enrico frowned. "Ouch!" he exclaimed. "Wasn't there *anything* you liked about my story?" he scoffed.

Grandfather laughed loudly. "I told you it's tough to hear criticism regarding your writing. Writing is a process and, as a writer, your work must undergo many revisions. This is normal, Enrico."

"Maybe I don't have the talent to become a published writer," Enrico confessed, doubting himself.

"You definitely have the talent, but do you have the perseverance? Great writers do not give up. They keep on revising until they get it right."

Enrico smiled. "I hear what you're saying and I won't give up. I'm ready to try again, and I will keep trying until my work is as good as it can be," Enrico said. Grandfather shook his hand. "That's my boy," he said. "I am very proud of you."

GO ON TO THE NEXT PAGE

Questions 39–46

39 Which pair of words BEST describes Grandfather?

   A   devious but kind

   B   tough but helpful

   C   mean but honest

   D   impatient but smart

40 What is the MAIN reason Grandfather does not want to read Enrico's story?

   A   He does not think it will be any good.

   B   He does not enjoy suggesting changes.

   C   He does not want to hurt Enrico's feelings.

   D   He does not believe Enrico will listen to him.

41 The following sentence appears in the first paragraph of the story.

"No, no, Enrico, don't do that," his grandfather gently chided.

Based on how the word is used in the story, which of the following BEST describes the meaning of "chided"?

   A   claimed

   B   corrected

   C   concealed

   D   challenged

GO ON TO THE NEXT PAGE

42 Which sentence BEST summarizes what happens in "Ouch!"?

A A boy tells his grandfather that he has decided to become a writer.

B A boy has learned a great deal from his grandfather over the years.

C A boy learns an important lesson about writing from his grandfather.

D A boy writes a story about a basketball player who lacks skill and size.

43 The boy in Enrico's story was often teased because

A he was shy.

B he played basketball.

C he liked to read.

D he was small.

44 What is one problem with Enrico's story?

A The outcome is obvious.

B The first paragraph is too long.

C The main character is ordinary.

D The language needs improvement.

45 Enrico asked Grandfather for help with his story because Grandfather

A learned English as a second language.

B has read many novels.

C has written several books.

D played basketball when he was younger.

GO ON TO THE NEXT PAGE

46 The author wants to reader to think that Enrico is

A   very talented.

B   very confused.

C   willing to work hard.

D   difficult to work with.

STOP

PART 4—WRITING

PART 4A: WRITING FROM KNOWLEDGE AND EXPERIENCE

DIRECTIONS:

In Part 4A, you will be given a theme and a number of ways to write about it. You must choose ONLY ONE way.

We will begin Part 4A together by reading the information on the next page. As I read aloud, please follow along on the page.

WAIT. DO NOT GO ON UNTIL TOLD TO DO SO.

Writing 4A: WRITING FROM KNOWLEDGE AND EXPERIENCE

47   THE IMPORTANCE OF FREEDOM

Freedom is the condition of being free. When you are free, you have the power to speak or think however you wish. Write about the importance of freedom.

Do **ONLY ONE** of the following:

describe a situation in which you are someone you know fought for freedom

OR

define freedom as it has affected your life

OR

explain why you think it is important to be free

OR

persuade the reader that freedom is worth fighting for

OR

write about the theme in your own way.

You may use examples from real life, from what you read or watch, or from your imagination.

Your audience will be interested adults.

You may use pages 208–209 for writing down ideas, organizing your thoughts, or writing a rough draft. Use the checklists on page 207 to help you improve your writing. Pages 236 and 237 contain the scorepoint descriptions used by readers to score your writing.

The final copy of your response must be written in the lined spaces on pages 222 and 223. Only the writing on this page will be scored. Nothing written on the prewriting and rough draft pages will be scored.

GO ON TO THE NEXT PAGE

CHECKLIST FOR REVISING AND PROOFREADING

DIRECTIONS:

Use the following checklists as you revise and proofread the writing you have done for Part 4A. When you are finished revising, you must write your final copy. Then, proofread your final copy to make sure that all of your revisions have been made.

CHECKLIST FOR REVISION:

_____ Do I have a clear central idea that connects to the theme?

_____ Do I stay focused on the theme?

_____ Do I support my central idea with important details/examples?

_____ Do I need to take out details/examples that DO NOT support my central idea?

_____ Is my writing organized and complete?

_____ Do I use a variety of words, phrases, and/or sentences?

CHECKLIST FOR EDITING:

_____ Have I checked and corrected my spelling to help readers understand my writing?

_____ Have I checked and corrected my punctuation and capitalization to help readers understand my writing?

CHECKLIST FOR PROOFREADING:

_____ Is everything in my final copy the way that I want it?

GO ON TO THE NEXT PAGE

NOTES/PLANNING

NOTES/PLANNING

PART 4B: STUDENT WRITING SAMPLE

DIRECTIONS:

In Part 4B, you will read a student writing sample, answer the multiple-choice questions that follow, and then write a short response. You may look back at the student writing sample as often as needed during Part 4B.

You may underline, highlight, or write notes in this booklet, but you must mark all of your answers in Part 4B of your **Answer Sheets** on page 224.

You may not use any resource materials (dictionaries, grammar books, spelling books, etc.) for any part of this test.

When you have finished Part 4B, STOP.

WAIT. DO NOT GO ON UNTIL TOLD TO DO SO.

DIRECTIONS: Read the following passage and answer the questions that follow. **You will have about 30 minutes to complete this section.**

Mr. Diaz

(1)   I would like to tell you a story about a teacher, the best teacher I have ever
(2) had, Mr. Diaz. I first had Mr. Diaz for English in the fifth grade. I am very shy and Mr.
(3) Diaz picked up on this right away. He did not call on me unless I raise my hand—and
(4) I didn't, for a real long time. Even though he didn't call on me, he would be sure to
(5) make eye contact with me. He knew that I was paying attention and following along.
(6)   While Mr. Diaz was very nice, he was also an amazing teacher. The class
(7) moaned at learning poetry, Mr. Diaz found poetry that was also the lyrics to some
(8) popular songs. He red the poems to us, but he also let us "listen" to them. He taught
(9) us about poets' lives—and some of them did not have it easy at all. It's easier to un-
(10) derstand poetry when you know where a poet is coming from.
(11)   Many other people also admire Mr. Diaz. Students have nominated him for
(12) many awards, but so far the more experienced teachers have won. In time—I am
(13) certain—Mr. Diaz will get the recognition he deserves.

Questions 48–52

48 In what line does the author use adjectives to describe Mr. Diaz?

   A   Line 12

   B   Line 2

   C   Line 9

   D   Line 6

49 What type of genre is reflected in this piece?

   A   memoir

   B   poetry

   C   legend

   D   biography

GO ON TO THE NEXT PAGE

50 All of the following ways to punctuate this sentence are correct EXCEPT:

    A   The class moaned at learning poetry, Mr. Diaz found poetry that was also the lyrics to some popular songs.

    B   When the class moaned at learning poetry, Mr. Diaz found poetry that was also the lyrics to some popular songs.

    C   The class moaned at learning poetry, so Mr. Diaz found poetry that was also the lyrics to some popular songs.

    D   The class moaned at learning poetry. Mr. Diaz found poetry that was also the lyrics to some popular songs.

51 What is the correct word choice for <u>raise</u> (line 3)?

    A   raising

    B   rose

    C   raised

    D   rosing

52 What is the correct spelling of the word <u>red</u> (line 8)?

    A   reed

    B   read

    C   rede

    D   rad

PEER RESPONSE TO THE STUDENT WRITING SAMPLE

DIRECTIONS: Write a response to the question in the box below. You may look back at the student writing sample as often as needed during Part 4B.

---

53 **Does the author convince you that Mr. Diaz is a great teacher? Why or why not?**

Use details from the student writing sample to support your answer.

Use the checklist on the next page to help you with your response. You may use the Notes/Planning space for writing down and organizing your ideas. Your response must be written in the lined spaces starting on page 225 of your **Answer Sheets**. Only the writing on your **Answer Sheets** will be scored. No extra sheets may be used.

You may not use any resource materials (dictionaries, grammar books, spelling books, etc.) for any part of this test.

When you have finished Part 4B, STOP.

GO ON TO THE NEXT PAGE

PART 4B: CHECKLIST FOR THE
PEER RESPONSE TO THE STUDENT WRITING SAMPLE

DIRECTIONS:

Use this checklist to help you with your response. Your response must be written in the lined spaces starting on page 225 of your **Answer Sheets**.

CHECKLIST:

_____   Do I clearly answer the question that was asked?

_____   Do I support my answer with details from the student writing sample?

_____   Is my response complete?

NOTES/PLANNING

GO ON TO THE NEXT PAGE

NOTES/PLANNING

PART 1A: PAIRED READING SELECTIONS

MARKING INSTRUCTIONS

Make heavy BLACK marks.
Erase cleanly.
Make no stray marks.

● **CORRECT MARK**   ◉ ⊘ ⊗ ◖ **INCORRECT MARK**

Multiple-choice questions

1. Ⓐ Ⓑ Ⓒ Ⓓ

2. Ⓐ Ⓑ Ⓒ Ⓓ

3. Ⓐ Ⓑ Ⓒ Ⓓ

4. Ⓐ Ⓑ Ⓒ Ⓓ

5. Ⓐ Ⓑ Ⓒ Ⓓ

6. Ⓐ Ⓑ Ⓒ Ⓓ

7. Ⓐ Ⓑ Ⓒ Ⓓ

8. Ⓐ Ⓑ Ⓒ Ⓓ

9. Ⓐ Ⓑ Ⓒ Ⓓ

10. Ⓐ Ⓑ Ⓒ Ⓓ

11. Ⓐ Ⓑ Ⓒ Ⓓ

12. Ⓐ Ⓑ Ⓒ Ⓓ

13. Ⓐ Ⓑ Ⓒ Ⓓ

14. Ⓐ Ⓑ Ⓒ Ⓓ

15. Ⓐ Ⓑ Ⓒ Ⓓ

16. Ⓐ Ⓑ Ⓒ Ⓓ

17. Ⓐ Ⓑ Ⓒ Ⓓ

18. Ⓐ Ⓑ Ⓒ Ⓓ

19. Ⓐ Ⓑ Ⓒ Ⓓ

20. Ⓐ Ⓑ Ⓒ Ⓓ

21. Ⓐ Ⓑ Ⓒ Ⓓ

Student Name_____

PART 1B: RESPONSE TO THE PAIRED READING SELECTIONS

Write your final response for question 22 here.

Write your final response for question 22 here.

PART 2: INDEPENDENT READING SELECTIONS

MARKING INSTRUCTIONS

Make heavy BLACK marks.
Erase cleanly.
Make no stray marks.

● CORRECT MARK   ◉ ⊘ ⊗ ◖ INCORRECT MARK

Multiple-choice questions

23. Ⓐ Ⓑ Ⓒ Ⓓ   31. Ⓐ Ⓑ Ⓒ Ⓓ

24. Ⓐ Ⓑ Ⓒ Ⓓ   32. Ⓐ Ⓑ Ⓒ Ⓓ

25. Ⓐ Ⓑ Ⓒ Ⓓ   33. Ⓐ Ⓑ Ⓒ Ⓓ

26. Ⓐ Ⓑ Ⓒ Ⓓ   34. Ⓐ Ⓑ Ⓒ Ⓓ

27. Ⓐ Ⓑ Ⓒ Ⓓ   35. Ⓐ Ⓑ Ⓒ Ⓓ

28. Ⓐ Ⓑ Ⓒ Ⓓ   36. Ⓐ Ⓑ Ⓒ Ⓓ

29. Ⓐ Ⓑ Ⓒ Ⓓ   37. Ⓐ Ⓑ Ⓒ Ⓓ

30. Ⓐ Ⓑ Ⓒ Ⓓ   38. Ⓐ Ⓑ Ⓒ Ⓓ

Student Name _____

PART 3: LINKED SELECTION

MARKING INSTRUCTIONS

Make heavy BLACK marks.
Erase cleanly.
Make no stray marks.

CORRECT
MARK

INCORRECT
MARK

Multiple-choice questions

39. Ⓐ Ⓑ Ⓒ Ⓓ

40. Ⓐ Ⓑ Ⓒ Ⓓ

41. Ⓐ Ⓑ Ⓒ Ⓓ

42. Ⓐ Ⓑ Ⓒ Ⓓ

43. Ⓐ Ⓑ Ⓒ Ⓓ

44. Ⓐ Ⓑ Ⓒ Ⓓ

45. Ⓐ Ⓑ Ⓒ Ⓓ

46. Ⓐ Ⓑ Ⓒ Ⓓ

Student Name_____

PART 4A: WRITING FROM KNOWLEDGE AND EXPERIENCE

Write your final response for question 47 here.

Write your final response for question 47 here.

PART 4B: STUDENT WRITING SAMPLE

MARKING INSTRUCTIONS

Make heavy BLACK marks.
Erase cleanly.
Make no stray marks.

● | ⊙ ⊘ ⊗ ◖
CORRECT | INCORRECT
MARK | MARK

Multiple-choice questions

48. Ⓐ Ⓑ Ⓒ Ⓓ

49. Ⓐ Ⓑ Ⓒ Ⓓ

50. Ⓐ Ⓑ Ⓒ Ⓓ

51. Ⓐ Ⓑ Ⓒ Ⓓ

52. Ⓐ Ⓑ Ⓒ Ⓓ

Student Name_____

Write your final response for question 53 here.

Write your final response for question 53 here.

ANSWER KEY

Posttest Answers

PART 1—READING

Part 1A: Paired Reading Selections

1   C   R.CS.08.01   Critical Standards
The author's main purpose was to inform readers about the life of a brave woman. Answer choice C is the correct answer. While the author does discuss aspects of slavery and fighting for what you believe in, the main focus is on the life of Sojourner Truth.

2   C   R.CS.08.01   Critical Standards
This article contains information on several topics, including the Civil War, a great speech, and the women's suffrage movement. However, it contains the *most* information about a great civil rights leader.

3   A   R.WS.08.07   Word Study
Isabella's victory in court was the first time she ever felt powerful. That is the meaning of the figurative phrase "her first taste of empowerment."

4   D   R.CM.08.01   Comprehension
The article explains that a sojourner is a traveler or a temporary resident of a place. Of these answer choices, a sojourner would most likely visit with the people in the town he or she had gone to.

5   D   R.CS.08.01   Critical Standards
When Sojourner repeated her question, she was trying to make her point clear. Hearing the question again and again kept reminding the audience of her purpose.

6   A   R.CM.08.02   Comprehension
Isabella became a traveling preacher after she changed her name to Sojourner Truth. In fact, the name Sojourner Truth was a way of signifying that she would now serve as a traveling preacher.

7   A   R.IT.08.02   Informational Text
One of the characteristics that made Sojourner Truth different from most other women (and men, too) was that she was very tall. She was about six feet tall, which was big for the time. This is a minor detail, but it is important in building a full understanding of why Sojourner was remarkable.

8   **D**   **R.NT.08.04   Narrative Text**
Sojourner's words create a mood of inspiration. She lists the many hardships and problems that she faced in her life, but points out that she is as strong as any man or woman. Answer choice D is the correct answer.

9   **C**   **R.CM.08.01 Comprehension**
Though slavery was a cause of the Civil War, this passage focuses on slavery during the time of the American Revolution. Answer choice C is correct.

10   **A**   **R.IT.08.02   Informational Text**
Lincoln was called the "Great Emancipator" because he declared the end of slavery in America during the Civil War.

11   **D**   **R.IT.08.02   Informational Text**
Most of the article discusses the history of slavery in chronological order, beginning in the 1600s and ending shortly after the American Revolution.

12   **B**   **R.IT.08.01   Informational Text**
The purpose of this passage is to inform readers about the slavery that existed in the northern states of the United States before the American Revolution. Answer choice B is the best answer.

13   **C**   **R.CS.08.01   Critical Standards**
The passage states that the final end to slavery in the North came during the battles of the Revolutionary War when both American and British troops promised slaves their freedom for fighting in the war.

14   **A**   **R.CM.08.03 Comprehension**
According to the third paragraph, in the 1600s, many slaves arrived in America as indentured servants. However, by the 1700s, slavery had been legalized, and slaves were treated as pieces of property.

15   **B**   **R.CM.08.02 Comprehension**
According to the fifth paragraph, America's Founding Fathers realized that it was ironic that they were fighting for a society that treated all people equally, yet they continued to hold slaves. They realized that slaves wanted the same freedom that they wanted.

16   **B**   **R.IT.08.03   Informational Text**
By agreeing to participate in the war, slaves were promised their freedom by both American and British forces.

17   **B**   **R.CM.08.03 Comprehension**
The only answer choice that applies to both passages is that slaves were forced to work hark and faced many hardships in their lives. Answer choice B is the best answer.

18   **A**   **R.CM.08.03** Comprehension

Both of these passages include information about the history of slavery in the United States. Sojourner Truth was a slave and spoke out against slavery. Likewise, "The First Emancipation" discusses slavery in the northern states prior to the American Revolution. Answer choice A is the best answer.

19   **D**   **R.CM.08.03** Comprehension

Only the Sojourner Truth passage discusses how women fought for the right to vote. Only "The First Emancipation" discusses how American changed after the Revolution. Answer choice C seems like it could work; however, Abraham Lincoln did not give his Emancipation Proclamation speech until the Civil War, and this passage focuses on the time surrounding the American Revolution. Answer choice D is the best answer.

20   **C**   **R.IT.08.02** Informational Text

From the passages, we don't learn much about Abraham's Lincoln's life, physical characteristics, or position on women's suffrage. However, we do know that both Sojourner Truth and Abraham Lincoln spoke out against slavery. Answer choice C is the best answer.

21   **B**   **R.CM.08.01** Comprehension

Both "The First Emancipation" and "Sojourner Truth: Traveling Preacher" send the message that all people, whether black, white, male, or female, have a right to be free and be treated equally. Answer choice B is the best answer.

Part 1B: Response to the Paired Reading Selections

W.PR. 08.01, .02, .03, .04, .05	**Writing Process**
W.GR.08.01	**Grammar and Usage**
W.SP.08.01	**Spelling**
W.HW.08.01	**Handwriting**

Michigan Educational Assessment Program (MEAP)
Grades 3–8

Rubric for the Response to the Paired Reading Selections

6	The student clearly and effectively chooses key or important ideas from each reading selection to support a position on the question and to make a clear connection between the reading selections. The position and connection are thoroughly developed with appropriate examples and details. There are no misconceptions about the reading selections. There are strong relationships among ideas. Mastery of language use and writing conventions contributes to the effect of the response.

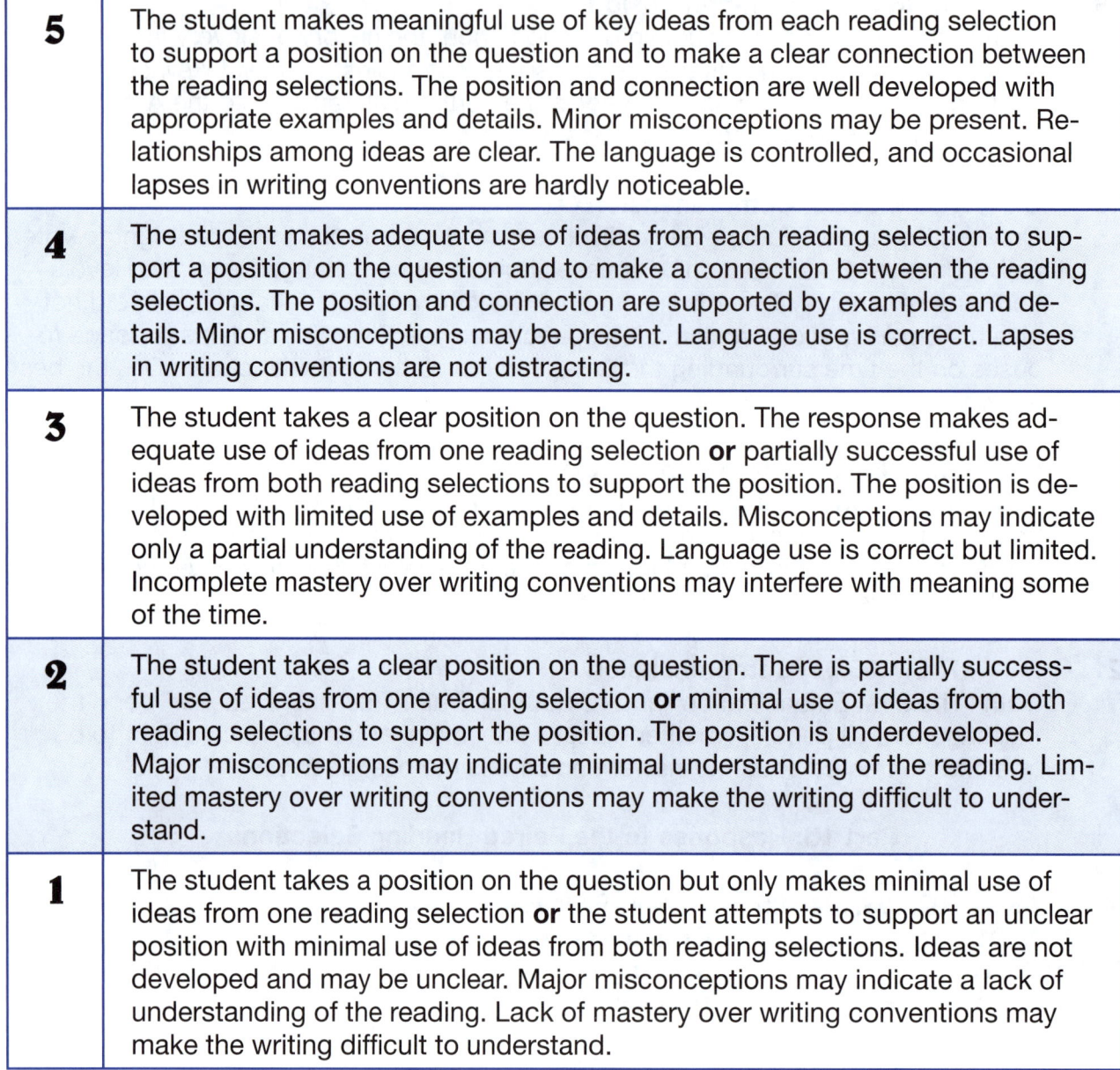

5	The student makes meaningful use of key ideas from each reading selection to support a position on the question and to make a clear connection between the reading selections. The position and connection are well developed with appropriate examples and details. Minor misconceptions may be present. Relationships among ideas are clear. The language is controlled, and occasional lapses in writing conventions are hardly noticeable.
4	The student makes adequate use of ideas from each reading selection to support a position on the question and to make a connection between the reading selections. The position and connection are supported by examples and details. Minor misconceptions may be present. Language use is correct. Lapses in writing conventions are not distracting.
3	The student takes a clear position on the question. The response makes adequate use of ideas from one reading selection **or** partially successful use of ideas from both reading selections to support the position. The position is developed with limited use of examples and details. Misconceptions may indicate only a partial understanding of the reading. Language use is correct but limited. Incomplete mastery over writing conventions may interfere with meaning some of the time.
2	The student takes a clear position on the question. There is partially successful use of ideas from one reading selection **or** minimal use of ideas from both reading selections to support the position. The position is underdeveloped. Major misconceptions may indicate minimal understanding of the reading. Limited mastery over writing conventions may make the writing difficult to understand.
1	The student takes a position on the question but only makes minimal use of ideas from one reading selection **or** the student attempts to support an unclear position with minimal use of ideas from both reading selections. Ideas are not developed and may be unclear. Major misconceptions may indicate a lack of understanding of the reading. Lack of mastery over writing conventions may make the writing difficult to understand.

22   The Trouble with Absolute Freedom

Sample 6-point answer: While freedom itself is a wonderful thing, absolute freedom would create chaos and give people permission to harm each other. If the President of the United States were to give people absolute freedom, we would not need laws at all because no one would have to follow them. Absolute freedom means doing whatever you want.

If absolute freedom existed in Sojourner Truth's time, she might be harmed by those around her who disagreed with her views on equality for women and blacks. If people were free to do as they pleased, they would be free to harm whomever they pleased. This is the primary danger of absolute freedom.

In "The First Emancipation" the author explains that Lincoln has been called "The Great Emancipator" for making slavery illegal. If absolute freedom existed, people might force others into slavery just because slaves are a source of free labor. As the author explains, many other emancipations occurred before Lincoln's time. If laws didn't exist and people were free to do as they pleased, some of these people fighting for the demolition of slavery might have been harmed or even killed.

PART 2—READING

Part 2: Independent Reading Selections

23　　**C**　　**R.WS.08.02　Word Recognition and Word Study**
To answer this question, the reader should look at the sentence before and the paragraph that follows the spot where the word "marooned" is found. In the previous sentence, Jim asks if Ben was shipwrecked. When he says that he was marooned, Jim recalls hearing the word and knows that it means that Ben had been deserted (choice C) by the rest of his crew on the island. The other choices don't fit the word as it is explained in the excerpt.

24　　**A**　　**R.NT.08.03　Narrative Text**
At the beginning of the excerpt, Jim is obviously scared, but once he realizes that Ben is not going to harm him, he is curious to learn more about how Ben ended up alone on the island. The other answer choices do not describe how Jim feels.

25　　**B**　　**R.CM.08.02　Comprehension**
The answer to this question can be found in the third paragraph of the excerpt. Once Jim remembers that he has something to defend himself with, he regains his courage and confronts Ben.

26　　**B**　　**R.NT.08.04　Narrative Text**
Here the reader must analyze the author's purpose. The best choice is answer B. Ben's appearance is a stark contrast to that of Jim and, as we are told, to any man that Jim has ever seen. This shows the reader the differences between the wilderness that Ben has had to live in and the civilized world that Jim comes from.

27　　**D**　　**R.WS.08.01　Word Recognition and Word Study**
Any of these choices are acceptable definitions of the word "sore," but by looking at the sentence, the reader should be able to see that the only choice that fits in terms of context is choice D. "Aching" could easily replace "sore" in the sentence, and is the correct choice.

28　　**A**　　**R.NT.08.01　Narrative Text**
Choices B, C, and D are all possible narrative genre categories that the excerpt could fall under. The only choice that does not fit is choice A, because the reader can clearly see that this is not a poem.

29   D   R.NT.08.04   Narrative Text
The answer to this question can be found in the beginning of the excerpt. Jim is saying that the fears he is familiar with, such as his fear of Silver, are preferable to the fears of the unknown, which is Ben.

30   C   R.CM.08.02 Comprehension
At the end of the excerpt, Jim says that he supposes Ben has gone crazy "in his solitude." The other options may also be true, but they are not the main reason that Jim feels Ben has gone crazy.

31   C   R.CM.08.02 Comprehension
The title of the article gives tells the reader the main idea of the article. Choices A, B, and D are all details included in the article, but choice C describes the main idea of the article.

32   A   R.CM.08.02 Comprehension
The correct answer to the question is choice A. Though the other choices may be true, the main reason the Sumerians built burial tombs was that they believed that their kings were gods that deserved this type of burial.

33   C   R.WS.08.01 Word Recognition and Word Study
The correct answer to this question is choice C. Choices A and D do not fit the meaning of the word "demise." Though "demise" can also mean "failure," the word that best fits the meaning of the word "demise" as used in the article is "death."

34   C   R.IT.08.03   Informational Text
Choices A and D do not describe the author's intent in writing this article. Choice B is partially true, but the main purpose of writing this article is to inform the reader about the ancient discovery of the Royal Cemetery at Ur.

35   D   R.IT.08.01   Informational Text
Choice A is not the correct answer, because the word "autobiography" means that the person that the story is about is also the author, meaning that Sir Leonard Woolley would have to have written it. The information in this article is a little too specific to be used in a brochure for a class on ancient civilizations (choice B) or an essay encouraging people to visit their local museums (choice C). The information would be best used in a research report on the Sumerians, because it tells the reader so much about their rituals and beliefs.

36   B   R.IT.08.03   Informational Text
Choice B is the correct answer. The author is not trying to show how fashion has changed (choice A). Choice C is not correct, because only a member of royalty would wear something so elaborate. Though the ornate jewelry does show that the Sumerians were advanced (choice D), the main reason that the author included this picture was to show the reader an example of the riches discovered by Woolley and his team.

37   C   R.CM.08.02   Comprehension
The final paragraph tells the reader that Woolley and his team found a gold cup near some of the burial party and this caused them to suspect that they drank poison so they could be with their queen.

38   B   R.IT.08.02   Informational Text
This question asks the reader to look at how the paragraph is organized and pick out the primary focus of the paragraph. The second paragraph talks about the beliefs of the common Sumerian people (choice B). The royal tombs are mentioned, but they are not the primary focus (choice C). Queen Puabi's tomb is talked about later in the article (choice A). The importance of Sumerian kings is also briefly mentioned, but the beliefs and customs of the people is the main idea of the second paragraph.

PART 3—READING

Part 3: Linked Selection

39   B   R.NT.08.03   Narrative Text
The best answer choice is B: Grandfather is tough; he wants Enrico to do his best, but he is also very helpful. Answer choice A is not correct, because there is no evidence that Grandfather is devious. Answer choice C is not correct, because Grandfather is not mean to Enrico, and answer choice D is not the best answer, because there really isn't any evidence that Grandfather is impatient.

40   C   R.NT.08.03   Narrative Text
In the story, Grandfather tells Enrico that he does not want to chance hurting Enrico's feelings. This is why he is reluctant to read the story. None of the other answer choices are supported by the story.

41   B   R.WS.08.01   Word Study
When the story says Grandfather gently chided Enrico, he has just corrected him. "Claimed" is not the same as "chided"; "concealed" would mean Grandfather has hidden something from Enrico, and Grandfather has not challenged Enrico, so answer choice B is the best answer.

42   C   R.CM.08.02   Comprehension
This question asks you to identify the best summary to the story. The summary is what the whole story is about. It is the main message of the story. Answer choice C is the best summary of the story, because it states that Enrico learned a lesson about writing. The other answer choices state main ideas.

43   D   R.NT.08.03   Narrative Text
In the story that the Enrico wrote, Miguel is teased because he is small and lacks skill while playing basketball. Answer choice D is the correct answer.

44 **A** **R.NT.08.03** **Narrative Text**
Grandfather tells Enrico that he knows what will happen in his story after reading only the first paragraph.

45 **C** **R.NT.08.03** **Narrative Text**
The story says that Grandfather is a writer and also worked as a reporter for a newspaper. This is the main reason Enrico asks him for help. Answer choice C is correct.

46 **C** **R.NT.08.03** **Narrative Text**
Although Enrico is at first frustrated by Grandfather's criticism, he later understands it can benefit his next writing attempts. So, instead of giving up, he's proven that he's willing to work hard to improve himself.

PART 4—WRITING

Part 4A: Writing from Knowledge and Experience

Michigan Educational Assessment Program (MEAP)

W.PR. 08.01, .02, .03, .04, .05	**Writing Process**
W.GR.08.01	**Grammar and Usage**
W.SP.08.01	**Spelling**
W.HW.08.01	**Handwriting**

Writing from Knowledge and Experience
Grades 3–8

Holistic Scorepoint Descriptions

6	The writing is exceptionally clear and focused. Ideas and content are thoroughly developed with relevant details and examples where appropriate. The writer's control over organization and the connections between ideas moves the reader smoothly and naturally through the text. The writer shows a mature command of language including precise word choice that results in a compelling piece of writing. Tight control over language use and mastery of writing conventions contribute to the effect of the response.

5	The writing is clear and focused. Ideas and content are well developed with relevant details and examples where appropriate. The writer's control over organization and the connections between ideas effectively moves the reader through the text. The writer shows a command of language including precise word choice. The language is well controlled, and occasional lapses in writing conventions are hardly noticeable.
4	The writing is generally clear and focused. Ideas and content are developed with relevant details and examples where appropriate, although there may be some unevenness. The response is generally coherent, and its organization is functional. The writer's command of language, including word choice, supports meaning. Lapses in writing conventions are not distracting.
3	The writing is somewhat clear and focused. Ideas and content are developed with limited or partially successful use of examples and details. There may be evidence of an organizational structure, but it may be artificial or ineffective. Incomplete mastery over writing conventions and language use may interfere with meaning some of the time. Vocabulary may be basic.
2	The writing is only occasionally clear and focused. Ideas and content are underdeveloped. There may be little evidence of organizational structure. Vocabulary may be limited. Limited control over writing conventions may make the writing difficult to understand.
1	The writing is generally unclear and unfocused. Ideas and content are not developed or connected. There may be no noticeable organizational structure. Lack of control over writing conventions may make the writing difficult to understand.

47   The Importance of Freedom

Sample 6-point response: My dad had a younger brother, who would have been my Uncle Mike. While my uncle Mike is no longer with us, I enjoy listening to stories about him and how he fought for freedom.

My uncle Mike was one of seven children. His father, my grandfather, was in the U.S. Army for many years. My grandfather fought in a war and he was proud to do so. He loves his country and thinks our freedom must be protected no matter what. Two of my dad's other brothers, my Uncle Brian and Uncle Dave, also enlisted in the service after high school. They served for a few years and then came home and went to college. My Uncle Mike was my grandparents' youngest child. My family says he was outgoing and popular. He played football and other sports and was very athletic.

Uncle Mike was a freshman in high school when an army recruiter visited their school. Uncle Mike told my grandfather that the recruiter was incredibly impressed with how much he knew about the U.S. Army. It was after talking to this recruiter that my uncle decided to go into the army. And, unlike my grandfather, Uncle Brian, and Uncle Dave, Uncle Mike wanted to make the army his career. He planned to serve for many, many years. Why did he want to do this? Uncle Mike had heard stories about people in other countries who were not free. These people did not live in a democracy. They could not choose what they wanted to be when they grew up. They were told what job they would do. It didn't matter if they liked this job or not. Uncle Mike knew this was wrong and decided to dedicate his life to protecting our freedom and helping people in other countries become free.

After high school, Uncle Mike enlisted in the service. He went to Georgia for basic training, which was difficult even for an accomplished athlete. But he persevered and did well. Then he flew to the Persian Gulf to work as a mechanic during the Gulf War. He returned home a few times after this and told my grandparents many stories about people's lives in the Middle East and how he hoped that these people would one day be free.

Uncle Mike was killed in an accident while fighting for freedom in the Persian Gulf. My grandparents were sad for a very long time, but they knew that Uncle Mike died doing what he loved: serving his country and helping others. Uncle Mike received several honors for his bravery. I, like the rest of my family, am very proud of him. Today my grandparents aren't as sad when they talk about Uncle Mike. We all like to remember him and talk about his life. We are proud that he fought for freedom.

Part 4B: Student Writing Sample

48   D   W.GR.08.01 Grammar and Usage
In line 6, the author says that Mr. Diaz was very nice and an amazing teacher. Answer choice D is correct.

49   A   W.GN.08.01 Writing Genres
This writing is most like a memoir. It's not a biography, a poem, or a legend.

50   A   W.GR. 08.01 Grammar and Usage
Independent clauses can be joined by a conjunction, a semicolon, or a period—but not a comma, as shown in answer choice A.

51   C   W.SP.08.01 Spelling
The author is writing in the past tense. "Raised" is the best answer.

52   B   W.SP.08.01 Spelling
While "red" is a correctly spelled word, the author should have used the word "read."

Michigan Educational Assessment Program (MEAP)

W.PR. 08.01, .02, .03, .04, .05	Writing Process
W.GR.08.01	Grammar and Usage
W.SP.08.01	Spelling
W.HW.08.01	Handwriting

Writing: Peer Response to a Student Writing Sample
Grades 3–8

Holistic Scorepoint Descriptions

Here is an explanation of what readers think about as they score your writing.

4	The response clearly and fully addresses the task and demonstrates an understanding of the effective elements of writing that are relevant to the task. Ideas are supported by relevant, specific details from the student writing sample. There may be surface feature errors, but they do not interfere with meaning.
3	The response addresses the task and demonstrates some understanding of the effective elements of writing that are relevant to the task. Ideas are somewhat supported with a mix of general and specific relevant details from the student writing sample. There may be surface feature errors, but they do not interfere with meaning.
2	The response demonstrates limited ability to address the task and may show limited understanding of the effective elements of writing that are relevant to the task. Ideas may be supported with vague and/or partially relevant details from the student writing sample. There may be surface features that partially interfere with meaning.
1	The response demonstrates an attempt to address the task with little, if any, understanding of the effective elements of writing that are relevant to the task. The response may include generalizations about the student writing sample with few, if any, details. There may be surface feature errors that interfere with meaning.

53

Sample 4-point response: The author definitely convinced me that, while Mr. Diaz lacks experience, he is a great teacher. The author says that Mr. Diaz was sensitive to his shyness, and didn't call on him unless he raised his hand. He also says that Mr. Diaz demonstrated that lyrics to songs are poetry, making his poetry lesson much more interesting. I would like to have a teacher like this.